Turing's Man

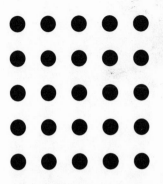

Turing's Man
Western Culture in the Computer Age

by J. David Bolter

The University of North Carolina Press

Chapel Hill

© 1984 The University of North Carolina Press
All rights reserved
Manufactured in the United States of America

First printing, March 1984
Second printing, June 1984

Library of Congress Cataloging in Publication Data

Bolter J. David, 1951–
   Turing's man.

   Bibliography: p.
   Includes index.
   1. Computers and civilization.   I. Title.
QA76.9.C66B64   1984   303.4′834   83-6942
ISBN 0-8078-1564-0
ISBN 0-8078-4108-0 pbk.

Contents

Figures

Preface

This book is a combination of the subject matter and outlooks of various disciplines. It seeks to explain a technical subject (digital computers and computer programming) in a historical light, calling upon the history of philosophical as well as technological ideas and going as far back as the Greeks. This seems an odd thing to do. It seems odd precisely because of that widely recognized and still dangerous split in our intellectual life—the gulf between the sciences and the humanities. Scientists, including applied scientists and engineers, know very little history or philosophy. Those in the humanities generally learn as little science as they can get away with. Many attempts have been made to bridge the gap, and this book can be read as one of them. I have chosen to write about computers because these machines should and, I think, will provide the sturdiest bridge between the world of science and the traditional worlds of philosophy, history, and art. The computer is a medium of communication as well as a scientific tool, and it can be used by humanists as well as scientists. It brings concepts of physics, mathematics, and logic into the humanist's world as no previous machine has done. Yet it can also serve to carry artistic and philosophical thinking into the scientific community. I am trying, in other words, to recognize and foster a process of cross-fertilization that has already begun.

In order to address scientists, engineers, and humanists, I must cover ground that is familiar to each group. My explanation of the computer is far too general to please the computer specialist, but I need to give the nonspecialist some idea of how the machine works in order to explain its impact on our culture. Readers with a background in classical and European philosophy and literature will probably quarrel with my many generalizations about "the Greek" or "the Western European" mind. Again I have to be general, for I need to introduce a wide range of topics in order to map out the areas of history and philosophy to which the computer is relevant. The reader may come to his own conclusions about the importance of pottery in the Greek world or that of the steam engine in the nineteenth century. He may argue that the idea of infinity is as important for philosophy and art today as it was a hundred years ago or that the notion of progress did exist in ancient times. My concern is not that the reader agree with all my conclusions but rather that he or she agree that it is important to think about computers from this perspective.

My premise is that technology is as much a part of classical and Western culture as philosophy and science and that these "high" and "lowly" expressions of culture are closely related. It makes sense to examine Plato and pottery *together* in order to understand the Greek world, Descartes and the mechanical clock together in order to understand Europe in the seventeenth and eighteenth centuries. In the same way, it makes sense to regard the computer as a technological paradigm for the science, the philosophy, even the art of the coming generation. Perhaps from this premise we can establish a much-needed dialogue among scientists, engineers, and humanists.

I wish to thank those who read my manuscript in various stages of preparation, particularly Dr. Phelps Gates, Dr. Philip Stadter (who provided encouragement and advice when both were needed), Mr. Peter Timmerman (whose comments helped to put me and keep me on track), and Dr. George Entenman (whose exacting criticism helped me tremendously in focusing my arguments). Thanks also go to the University of North Carolina Press and particularly to Mr. Lewis Bateman, for the patience and willingness to support this rather unusual project. In editing the manuscript, Ms. Pam Morrison has saved me from many inconsistencies and helped to make the arguments clearer and the prose more readable. Finally, I wish to thank my wife, Christine de Catanzaro, for her valuable criticism and boundless support at every phase of the writing of this book.

New Haven, October, 1982

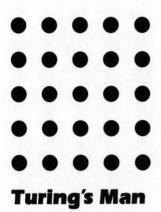

Turing's Man

1 Introduction

The Measure of Technological Change

We live in spectacular but very uncertain times. In some ways, the prospects for the future have never been more exciting: they include great advances in the physical sciences, the freeing of men and women from all dangerous and dreary work through automation, and the exploration of outer space. At the same time, the social and economic problems of the near future are staggering—enormous overpopulation, the scarcity of resources, and the deterioration of the environment. Many ages in the past have shown great promise while facing great difficulties, yet our age is perhaps unique in that its problems and its promise come from the same source, from the extraordinary achievements of science and technology. Other factors have remained constant. Men and women are no more greedy, violent, compassionate, wise, or foolish than before, but they find themselves in command of a technology that greatly enhances their capacity to express these all-too-human qualities. Technology enables them to reshape nature to conform to their needs, as far as possible to make nature over in their image.

But it is a flawed image. Mankind is indeed both good and evil, and it is to be expected that human technology will sometimes harm nature rather than improve it. Until recently, however, our technical skills were so feeble in comparison with the natural forces of climate and chemistry that we could not seriously affect our environment for good or ill, except over millennia. High technology promises to give us a new power over nature, both our own nature and that of our planet, so that the very future of civilization now depends upon our intelligent use of high technology.

There are many elements in our technological future, but the key, I think most will agree, is our burgeoning electronic technology. By focusing upon it, I hope to suggest possibilities and limitations that apply to the whole of our technological world. This essay is not a primer on computer programming, nor does it provide a thorough technical explanation of how computers work. It is instead a study of the impact that electronic logic machines are having upon our culture. It does not concern the immediate economic and political impact, although these are interesting in their own right. It does concern a subtler effect that is more difficult to describe but in the long run perhaps more important: a change in the way men and women in the electronic age think about themselves and the world around them.

It is true that the most visible results of technological change are economic. Automation—the Europeans call it "rationalization"—is altering every traditional industry, bringing increased production but also threatening jobs. The trend will likely continue in the coming decades, as microelectronics permits machines and machine tools to become "smart": to be programmed for several related tasks rather than the rigid repetition of one task. The file clerk's job is already in danger; for nearly three decades computers have been processing inventories, billing customers, and providing endless reports for major corporations, and minicomputers are now doing the same for smaller businesses. Whether all this information processing necessarily improves efficiency is not the question. Most large companies claim that they can no longer handle the volume of their business without electronic equipment.

This might now be said of Western society in general. Because there are so many people, demanding so many services from both the government and the private sector, computers have become indispensable. Critics of the computer age see dangerous consequences of this dependency. The social atmosphere in which we work and live is being poisoned by these new machines. The opportunities for human beings to respond humanely toward one another are lost when each is treated not as an individual with a unique history and unique problems but as an identification number to which is attached a vector of quantified data. As our society moves toward the conviction that there is nothing important in the human condition that cannot be quantified and fed as data into a digital computer, the positive qualities of Western humanism may well be lost.

Are the critics right? Does the computer threaten the values of a humane society? Electronic technology is not yet well enough understood to evaluate properly its potential for good or ill. Surely we do not explain how electronics is changing our society simply by pointing to its widespread economic impact. Merely to rehearse the number of jobs lost to automation, the number of bank transactions conducted electronically, or the amount of money invested in computers by corporations is to commit the same error that is charged to the computer itself—to mistake a quantitative analysis for a qualitative one. The economic background is important because the business world has financed the frenetic pace at which electronic technology has developed over the past thirty years, and today the large majority of machines is devoted to such mundane purposes as controlling inventories and preparing invoices.

Economic conditions have only served to make computers commonplace and therefore potentially influential: what is allowing them to realize this potential is not merely their numbers but their peculiar qualities as machines. The same was true of the automobile and the telephone. Until there were millions of cars on the road, they could not change the character of American society: allow the middle class new physical and social mobility, give writers a new, largely negative symbol of the American technological spirit, and so on. Before the telephone was common, it could not serve as a means of distance-annihilating communication for the millions and again for authors as a peculiar symbol of isolation and distance between men. The economic conditions simply enabled these devices to express qualities that were latent in them from the moment the first prototypes were tinkered together.

So it is with the computer, with the difference that the electronic technology has not yet reached the maturity that the revolutions in communications and transport achieved in America in the 1920s. Computers affect the lives of all members of North American and European society, but so far largely at secondhand. Banks use them to keep our accounts, the government to calculate our taxes and to take the census, but these expensive and arcane machines have remained out of personal reach, in the possession of organizations with the capital needed to buy them and the personnel to operate them. Most laymen have never been in the same room with a computer (except for electronic games and the ubiquitous pocket calculator). This will surely change when powerful microprocessors become available to the middle class at prices that make them increasingly attractive, first as toys and then as amenities of modern life. When computers too are counted in the millions, rather than the tens of thousands, they will manifest their full impact upon our society. In a matter of years, most educated people will be using computers in their work, and this is far more important than the fact that home computers will serve as entertainment centers or menu-planners.

We have already reached the stage in which physicists and chemists and many biologists call upon electronic logic to measure their experiments and to help interpret results, statistically or through models. Sociologists and economists cannot do without computers. Humanists, scholars, and creative writers have as yet little use for these machines. This too will change, rather rapidly, as they realize that at least text editing by computer is far easier than working with pens, paper, or typewriters. Even now, the publishing industry understands the advantages of electronically controlled photocomposition. Further in the future, we may well expect libraries of information, literary as well as scientific, to be stored on electric media and made available by computer. Whether literature, philosophy, or the study of history could ever be quantified and made into input for a program is not the question. Humanists as well as scientists will employ computers, simply because these devices will be a principal medium of communication for the educated community of Europe and North America. The philosophy and fiction of the next hundred years will be written at the keyboard of a computer terminal, edited by a program, and printed under electronic control—if indeed such works are printed at all, for they may simply be stored on magnetic disk and called up electronically by the reader. In the long run, the humanist will not be able to ignore the medium with which he too will work daily: it will shape his thought in subtle ways, suggest possibilities, and impose limitations, as does any other medium of communication.

Think of a woodcut or painting of a scholastic monk living in the late Middle Ages. We see the man dressed in a habit and crowded with his precious books into a small cell, and perhaps through the window we catch a glimpse of the grounds of his monastery. He sits or stands at a high inclined desk with one or two large volumes before him; perhaps he is composing a treatise or laboriously copying a manuscript. He works by candlelight or daylight, and on the wall behind him there hangs an astrolabe or a compass for geometry. It is not hard to imagine that every element in such a picture has its bearing upon the metaphysical as well as the everyday thinking of a medieval or Renaissance schoolman. The habit and the cell itself represent the social conditions under which scholastic thought flourished: their bearing is obvious. The stylus and indeed the alphabet with which he writes, the parchment he uses, the fact that he must work from manuscripts rather than printed books, the precious authority of the few works he possesses, the absence of reliable electric light, the quality of the scientific and other instruments at his disposal—all these elements too have an influence on his work. One picture cannot fully characterize a way of life, yet if we could step into one such cell, handle the books and tools, or walk around the grounds of a functioning monastery before the Reformation and the invention of printing, we would clearly be in a better position to understand the summae, biblical commentaries, treatises on logic, and collections of letters that have come down to us. In the same way, the ancient mosaic or wall painting of a Roman poet, reclining in his garden and composing to a literate slave, who takes down the lines on a wax tablet or roll of papyrus, has much to tell us about ancient literary and philosophical thinking and the particular genres that served to express that thinking.

The next archetypal picture will be a photograph of a scientist or philosopher seated at a computer terminal; in front of him will be a television screen displaying the words as he types. The room will be low-lit, because the words and diagrams on the screen will themselves be illuminated, and sparsely furnished, because most of the references and working tools will be in the computer itself. Memory devices will hold experimental results or literary texts; programs will copy texts and present results in legible formats. The blinking cursor on the screen, far more convenient than the medieval copyist's stylus, will erase errors; editing programs will be more responsive and careful than the scribe who took down the lines of the ancient poet. The scientist or philosopher who works with such electronic tools will think in different ways from those who have worked at ordinary desks with paper and pencil, with stylus and parchment, or with papyrus. He will choose different problems and be satisfied with different solutions.

The Computer as a Defining Technology

In the past, even a major new technology of materials or power has seldom done away with its predecessor entirely. Instead one technology relegates another to subservience, to tasks at which the new technology is either inappropriate or uneconomical. The invention of iron did not eliminate bronze tools, which were cheaper and easier to make, nor did effective windmills and waterwheels eliminate the use of harnessed animals, since there was no convenient way to pull a cart over land with wind or water power. The steam engine, the internal combustion engine, and now the nuclear reactor are unlikely to replace the workman using nothing more sophisticated than a dolly to get a heavy load up a short flight of stairs. We still rely today upon skills and discoveries (fire, farming, mining) that are thousands of years old; electronics is not the technology most important to our survival or prosperity. In that sense the Neolithic invention of agriculture has never been rivaled. We depend today upon a small number of farmers using high technology to feed us and so free us to ponder the significance of computers or anything else.

Computers perform no work themselves; they direct work. The technology of "command and control," as Norbert Wiener has aptly named it, is of little value without something to control, generally other machines whose function is to perform work. For example, the essence of the American space shuttle is the computers that control almost every phase of its operation. But unless the powerful rocket engines provide the expected thrust, there is no mission for the computers to control. The computer leaves intact many older technologies, particularly the technologies of power, and yet it puts them in a new perspective. With the appearance of a truly subtle machine like the computer, the old power machines (steam, gas, or rocket engines) lose something of their prestige. Power machines are no longer agents on their own, subject only to direct human intervention; now they must submit to the hegemony of the computer that coordinates their effects.

As a calculating engine, a machine that controls machines, the computer does occupy a special place in our cultural landscape. It

is the technology that more than any other defines our age. Our
generation perfected the computer, and we are intrigued by possibilities as yet only half-realized. Ruthlessly practical and efficient, the computer remains something fantastic. Its performance
astonishes even the engineers who build it, just as the clock must
have astonished craftsmen in the fourteenth century and the
power of the steam engine even the rugged entrepreneurs of the
nineteenth century. For us today, the computer constantly threatens to break out of the tiny corner of human affairs (scientific
measurement and business accounting) that it was built to occupy, to contribute instead to a general redefinition of certain
basic relationships: the relationship of science to technology, of
knowledge to technical power, and, in the broadest sense, of
mankind to the world of nature.

This process of redefinition is not new. Technology has always
exercised such an influence; it has always served both as a bridge
and a barrier between men and their natural environment. The
ability to make and use tools and the subtle capacity to communicate through language have allowed men to live more comfort-
ably in the world, but these achievements have also impressed
upon them their separation from nature.

Men and women throughout history have asked how it is that
they and their culture (their technology in the largest sense) transcend nature, what makes them characteristically human and not
merely animal. For the Greeks, a cardinal human quality was the
ability to establish the political and social order embodied in a
city-state: men at their best could set collective goals, make laws,
and obey them, and none of these could be achieved by animals
in a state of nature. Their city-state was a feat of social technology. In the Middle Ages, the accomplishments of technology
were perhaps more physical than social, but the use of inanimate
sources of power, wind and water, fostered a new view of mankind versus the forces of nature. The discoveries of the Renaissance and the Industrial Revolution moved men closer to nature
in some respects and separated them even more radically in
others. Continued emphasis on exploring and manipulating the
physical world led to a deeper appreciation of the world's resources. Yet the desire to master nature—to harness her more
efficient sources of power in steam and fossil fuels and to mine
her metals for synthetic purposes—grew steadily throughout this
period. When Darwin showed convincingly that man was an
animal like any other, he shattered once and for all the barrier

that separated men from the rest of nature in the Greek and medieval chains of being. Yet nineteenth-century engineers with their railroads and still more twentieth-century physicists with their atomic bombs seemed less natural than ever before, less under the control of either nature or a personal deity and more responsible for their own misjudgments.

Continually redrawing the line that divides nature and culture, men have always been inclined to explain the former in terms of the latter, to examine the world of nature through the lens of their own created human environment. So Greek philosophers used analogies from the crafts of pottery and woodworking to explain the creation of the universe: the stars, the planets, the earth, and its living inhabitants. In the same way, the weight-driven clock invented in the Middle Ages provided a new metaphor for both the regular movements of heavenly bodies and the beautifully intricate bodies of animals, whereas the widespread use of the steam engine in the nineteenth century brought to mind a different, more brutal aspect of the natural world. It is certainly not true that changing technology is solely responsible for mankind's changing views of nature, but clearly the technology of any age provides an attractive window through which thinkers can view both their physical and metaphysical worlds.

Technology has had this influence even upon philosophers, like Plato, who generally disdain human craftsmanship and see it as a poor reflection of a greater nonhuman reality. And even in Christian theology and poetry, the pleasures of heaven could only be described as grand versions of the tainted pleasures men know on earth, and the tortures of hell were earthly tortures intensified. Almost every sort of philosopher, theologian, or poet has needed an analogy on the human scale to clarify his or her ideas. Speaking of creation as the imposition of order upon the natural world, he or she generally assumes a creator as well, and this creator is a craftsman or technologist.

It is in this context that I propose to examine electronic technology. The computer is the contemporary analog of the clocks and steam engines of the previous six centuries; it is as important to us as the potter's wheel was to the ancient world. It is not that we cannot live without computers, but that we will be different people because we live with them. All techniques and devices have the potential to become defining technologies because all to some degree redefine our relationship to nature. In fact, only a few devices or crafts in any age deserve to be called defining technologies. In the ancient world, carpentry and masonry were about as important as spinning and pottery, and yet poets and philosophers found the latter two far more suggestive. In medieval Europe, crop rotation and the moldboard plough had a greater economic and social impact than the early clockwork mechanisms. Yet not many philosophers and theologians compared the world to a lentil bean. Certain skills and inventions have moved easily out of the agora into the Academy, out of the textile mill into the salon, or out of the industrial research park into the university classroom.

The vision of particular philosophers and poets is important to such a transference. Descartes and his followers helped to make the clock a defining technology in Western Europe. Certainly the first poet to elaborate the myth of the Fates who spin the thread of life helped to make textiles a defining technology for ancient Greece. But there must be something in the nature of the technology itself, so that its shape, its materials, its modes of operation appeal to the mind as well as to the hand of their age—for example, the pleasing rotary motion of the spindle or the autonomy and intricacy of the pendulum clock.

Such qualities combine with the social and economic importance of the device to make people think. Very often a device will take on a metaphoric significance and be compared in art and philosophy to some part of the animate or inanimate world. Plato compared the created universe to a spindle, Descartes thought of animals as clockwork mechanisms, and scientists in the nineteenth century and early twentieth centuries have regularly compared the universe to a heat engine that is slowly squandering its fuel. Today the computer is constantly serving as a metaphor for the human mind or brain: psychologists speak of the input and output, sometimes even the hardware and software, of the brain; linguists treat human language as if it were a programming code; and everyone speaks of making computers "think."

A defining technology develops links, metaphorical or otherwise, with a culture's science, philosophy, or literature; it is always available to serve as a metaphor, example, model, or symbol. A defining technology resembles a magnifying glass, which collects and focuses seemingly disparate ideas in a culture into one bright, sometimes piercing ray. Technology does not call forth major cultural changes by itself, but it does bring ideas into a new focus by explaining or exemplifying them in new ways to larger audiences. Descartes's notion of a mechanistic world that

obeyed the laws of mathematics was clear, accessible, and therefore powerful because his contemporaries lived with clocks and
gears. So today electronic technology gives a more catholic appeal to a number of trends in twentieth-century thought, particularly the notions of mathematical logic, structural linguistics, and
behavioral psychology. Separately these trends were minor up-
heavals in the history of ideas; taken together, they become a major revision in our thinking.

Turing's Man

In the development of the computer, theory preceded practice.
The manifesto of the new electronic order of things was a paper
("On Computable Numbers") published by the mathematician
and logician A. M. Turing in 1936. Turing set out the nature and
theoretical limitations of logic machines before a single fully programmable computer had been built. What Turing provided was a
symbolic description, revealing only the logical structure and
saying nothing about the realization of that structure (in relays,
vacuum tubes, or transistors). A Turing machine, as his description came to be called, exists only on paper as a set of specifications, but no computer built in the intervening half century has
surpassed these specifications; all have at most the computing
power of Turing machines. Turing is equally well known for a
very different kind of paper; in 1950 he published "Computing
Machinery and Intelligence." His 1936 work was a forbidding
forest of symbols and theorems, accessible only to specialists.
This later paper was a popular polemic, in which Turing stated
his conviction that computers were capable of imitating human
intelligence perfectly and that indeed they would do so by the
year 2000. This paper too has served as a manifesto for a group
of computer specialists dedicated to realizing Turing's claim by
creating what they call "artificial intelligence," a computer that
thinks.

Put aside for the moment the question of whether the computer
can ever rival human intelligence. The important point is that
Turing, a brilliant logician and a sober contributor to the advance
of electronic technology, believed it would and that many have
followed him in that belief. The explanation is partly enthusiasm
for a new invention. In 1950 the computer was just beginning to
bring vast areas of science and business under its technological

aegis. These machines were clearly taking up the duties of command and control that had always been assumed by human operators. Who could say then where the applications of electronic command and control might end? Was it not natural to believe that the machine would in time eliminate the human operator altogether? Inventors, like explorers, have a right to extravagant claims. Edison had said that the record player would revolutionize education; the same claim was made for radio and, of course, television.

I think, however, that Turing's claim has had a greater significance. Turing was not simply exaggerating the service his machine could perform. (Does a machine that imitates human beings perform any useful service at all? We are not running short of human beings.) He was instead explaining the meaning of the computer for our age. A defining technology defines or redefines man's role in relation to nature. By promising (or threatening) to replace man, the computer is giving us a new definition of man, as an "information processor," and of nature, as "information to be processed."

I call those who accept this view of man and nature Turing's men. I include in this group many who reject Turing's extreme prediction of an artificial intelligence by the year 2000. We are all liable to become Turing's men, if our work with the computer is intimate and prolonged and we come to think and speak in terms suggested by the machine. When the cognitive psychologist begins to study the mind's "algorithm for searching long-term memory," he has become Turing's man. So has the economist who draws up input-output diagrams of the nation's business, the sociologist who engages in "quantitative history," and the humanist who prepares a "key-word-in-context" concordance.

Turing's man is the most complete integration of humanity and technology, of artificer and artifact, in the history of the Western cultures. With him the tendency, implicit in all eras, to think "through" one's contemporary technology is carried to an extreme; for him the computer reflects, indeed imitates, the crucial human capacity of rational thinking. Here is the essence of Turing's belief in artificial intelligence. By making a machine think as a man, man recreates himself, defines himself as a machine. The scheme of making a human being through technology belongs to thousands of years of mythology and alchemy, but Turing and his followers have given it a new twist. In Greek mythology, in the story of Pygmalion and Galatea, the artifact, the perfect ivory statue, came to life to join its human creator. In the seventeenth and eighteenth centuries, some followers of Descartes first suggested crossing in the other direction, arguing, with La Mettrie, that men were no more than clockwork mechanisms. Men and women of the electronic age, with their desire to sweep along in the direction of technical change, are more sanguine than ever about becoming one with their electronic homunculus. They are indeed remaking themselves in the image of their technology, and it is their very zeal, their headlong rush, and their refusal to admit any reservation that calls forth such a violent reaction from their detractors. Why, the critics ask, are technologists so eager to throw away their freedom, dignity, and humanity for the sake of innovation?

Should we be repelled by the notion of man as computer? Not until we better understand what it means for man to be a computer. Why on the face of it should we be more upset by this notion than by the Cartesian view that man is a clock or the ancient view that he is a clay vessel animated by a divine breath? We need to know how Turing's man differs from that of Descartes or Plato, how the computer differs conceptually and symbolically from a clock or a clay pot. And to do this, we must isolate the precise qualities of computers and programming, hardware and software, that have the magnifying effect mentioned earlier— bringing ideas from philosophy and science into a new focus.

●　●　●　●　●
●　●　●　●　●
●　●　●　●　●
●　●　●　●　●
●　●　●　●

2　Defining Technologies in Western Culture

In the spring of 1900, a group of sponge divers came upon an ancient shipwreck off the island of Antikythera in the Aegean. Among the statues and objects recovered were "some calcified lumps of corroded bronze," as Derek J. de Solla Price puts it ("An Ancient Greek Computer," 61). These were recognized some months later as fragments of a mechanism, but it was not until 1955 that Price and his colleagues managed to sort out the pieces and reconstruct this remarkable machine. The Antikythera device is "like a great astronomical clock without an escapement, or like a modern analogue computer, which uses mechanical parts to save tedious calculations" ("An Ancient Greek Computer," 66). By turning a crank, the user would set in motion various gears and read on dials the position of the sun and the planets in the zodiac. The discovery of this piece, dated to the first century B.C., was a total surprise, proving that some craftsmen in the Greek world of the eastern Mediterranean were thinking in terms of the mechanization and mathematization of time and the heavens long before the Middle Ages, when such notions began to take hold of Western culture.

The Antikythera device (figure 2-1) may have been a masterpiece of ancient craftsmanship, but it was not a part of the defining technology of the ancient world. To quote Price again: "Nothing like this instrument is preserved elsewhere. Nothing comparable to it is known from any ancient scientific text or literary allusion. On the contrary, from all that we know of science and technology in the Hellenistic Age we should have felt that such a device could not exist" ("An Ancient Greek Computer," 60). Clearly there is much that we do not know about the Hellenistic Age; the real surprise is that such a remarkable invention should go unmentioned by technical writers and philosophers of the time. Even if, as Price believes, there was a tradition of making such mechanisms, which was handed on to Arab craftsmen, that tradition did not speak to its age. Poets compared the heavens to a giant spindle shining with fires, not to a clockwork mechanism. The Greeks may have devised the first analog computer, but the device did not change them into Cartesian or Turing's men.

In assessing past technology, there is great danger of anachronism. A device may look like a computer to our hindsight, yet it may not *be* a computer to contemporaries, that is, it may not function for its culture as a computer does today. It may serve instead as an ornament or a toy. On the other hand, our own technology cannot be assessed except in reference to the past. The cultural meaning of the computer becomes clear only in comparison with the meaning of the clock for its age and the steam engine for its age. Since our goal is to explore the fascination of Turing's man with his new machine, we must put this fascination in its historical context.

We cannot, and we need not, account for every major craft or machine from ancient Greece to the present; again what matters are those that caught the interest of contemporary thinkers—the defining technologies. Technologically and in other ways, civilizations proceed at their own pace and with their own inner logic, and each possesses a characteristic set of materials, techniques, and devices that help to shape its cultural outlook. For Greece and Rome, the materials included clay, wool, and wood; the devices, the drop spindle and the potter's wheel. The ancients had a manual technology. For Western Europe, metal and coal assumed a new prominence; new or radically improved devices included the waterwheel and windmill, the clock, and the steam engine. These devices characterized a mechanical-dynamic technology. In both cases philosophers or poets used contemporary technology to aid their flights of imagination. In the ancient world, they observed the drop spindle and potter's wheel and were struck by the use of rotary motion and of human or animal power. From these observations came support for the rotating universe, the animate nature of the stars, Aristotle's theory of form and matter. Western European thinkers found first the clock and then the steam engine more suggestive than the manual crafts still practiced in their day and were impressed by the qualities of mechanically translated and controlled motion. They spoke of a clockwork world in which even animals were intricately regulated machines, or they regarded the world itself as a gigantic, inefficient steam engine.

For understanding Turing's man, the craftsmanship of Greece and Rome (from the Bronze Age to about the fifth century after Christ) has almost as much to interest us as the later age of the machine in Western Europe and North America. Admittedly, the science and engineering of the Industrial Revolution (such as electrodynamic theory and machine tooling) provided the skills that have allowed our age to build the computer. Ancient science and craftsmanship made no such direct contribution. The experience of Greece and Rome is far more remote from ours, and in one sense its value lies in its remoteness. On the other hand, although the ancients were not especially innovative in their crafts and tools, their literature is marked by an imaginative reaction to their technology. Paradoxically, there are aspects of electronic technology leading us away from the thinking of the recent past and closer to the ancient world. The age of the computer is in some ways a return to the age of the potter's wheel.

Manual Technology and the Ancient World

Technology may be regarded largely as the controlled application of power. Whether the source is human muscle or nuclear energy, the idea is to channel power so as to modify some natural material in some useful way: to shape clay, to weave cloth, to purify metals and cast them into molds, or to manipulate electronic data. Ultimately, the guiding intelligence for any technical process is the man or woman who creates the process, but in modern times there have been a variety of self-regulating machines that control the immediate application of power. In the ancient world, however, there was very little machinery to mediate between the human craftsman and the materials with which he worked. The craftsman's hands, or a tool-in-hand, provided the control for most production, and human or animal muscle provided the power.

Take a thin rounded shaft, add a weight at the bottom and a hook at the top, and suspend the shaft by a loose roving of yarn. You have made an ancient *drop spindle*, a device for twisting yarn into thread. The weight at the bottom of the spindle, generally a disk or cup, was called a *whorl*. Give the spindle a turn, and the fibers in the yarn will be drawn into thread. By paying out the yarn skillfully from a distaff in one hand and shaping the fibers with the other, you can achieve a fine, even product. These two elements caught the ancient imagination: the rotating motion of the spindle and the handicraft involved in shaping the thread (see figure 2-1). Take a larger shaft of wood, put a wheel at the bottom as a weight, anchor the shaft in some sort of table, and join a flat terra-cotta disk to the top. Now you have an ancient potter's wheel, which was suggestive for just the same reasons as the drop spindle. Clay was placed on the disk, and the shaft was set spinning by a helper or by the potter himself. The potter gave shape to the turning clay beneath his hands, drawing the material out and allowing the wheel to provide the symmetry.

Everywhere we look in ancient technology, we see the same qualities: craftsmen working very close to their materials, imparting motion and supplying what little power they need by their own or animal muscle. The hand loom (vertical at first, with warp threads tied to potsherds) and the carpenter's lathe were simple devices but were used with considerable skill. Stone masons and architects operated with the same sense of proportion on a much larger scale and with a more intractable medium, shaping their limestone and marble blocks accurately with few power tools. Roman aqueducts sometimes ran for tens of miles, maintaining the proper grade and employing siphons or tiered bridges to span valleys. Such construction is more impressive when we remember that animals and men with winches and levers moved every block of stone into place.

In contrast to later Western Europeans, the Greeks and Romans did not show much interest in improving the sources of power available to them or in finding new sources. Slaves, animals, and craftsmen were the prime movers, and even these were not always used at their greatest efficiency. R. J. Forbes puts it simply: "During almost the whole of antiquity the only prime movers were men and animals. Even when the more economical form of the water-mill was devised about the beginning of the Christian era, its penetration was so slow that the situation remained practically unchanged for another 400 years" (Singer et

al., *History of Technology*, 2:590). As early as the first century
before Christ, the Romans designed an effective waterwheel, particularly useful for grinding grain, but it was not exploited for hundreds of years. Indeed, the Romans often used slaves in squirrel cages to raise water, rather than allowing falling water to work for men. The windmill was apparently unknown to the ancients. In short, there was something in the world outlook of the ancients (perhaps the reliance on slavery) that kept them satisfied with traditional sources of power and did not compel them, like later Europeans, to seek to increase efficiency, invent new prime movers, and in general expand their control and domination of nature.

The result was a simple but elegant technology of the hand rather than of the machine. The ancient craftsman worked with tools that became extensions of his hands in the manipulation of his materials. There was no real mass production. Although a pottery shop in Athens might employ seventy men who worked from specified designs, each thrown pot carried to some extent the impress of the hand that made it. Also, all technical discoveries were the product of clever observation and innovation without a theoretical basis, for the relationship between science and technology, so much a part of our own industrial society, did not exist.

With the possible exception of astronomy and mathematics, there were no theoretical sciences in the ancient world. The ironsmith did not go to a chemist to discover what sort of impurities were in his ore or what temperatures were needed to remove them. The Roman architect did not consult a physicist to compute the distribution of weight over one of his vaults. Advances came from educated trial and error, and the knowledge gained through such experience was passed on, as craft technology always is, as a series of rules of thumb, techniques for achieving results. In important crafts, such as architecture and warfare, there was some didactic literature. But most techniques—for pottery, carpentry, and the like—would be handed down by some system of apprenticing. The great "sciences" of antiquity with long literary traditions were philosophy, history and antiquarianism, and rhetorical and literary theory, and only a few men of learning and creative intelligence, notably Archimedes and Hero of Alexandria, were concerned with technology. Ancient technology was caught in this vicious circle: without theoretical foundations (ties to a legitimate physics and chemistry), its problems held little interest for ancient philosophers and

Figure 2-1. Antikythera Device and Spindle Forms

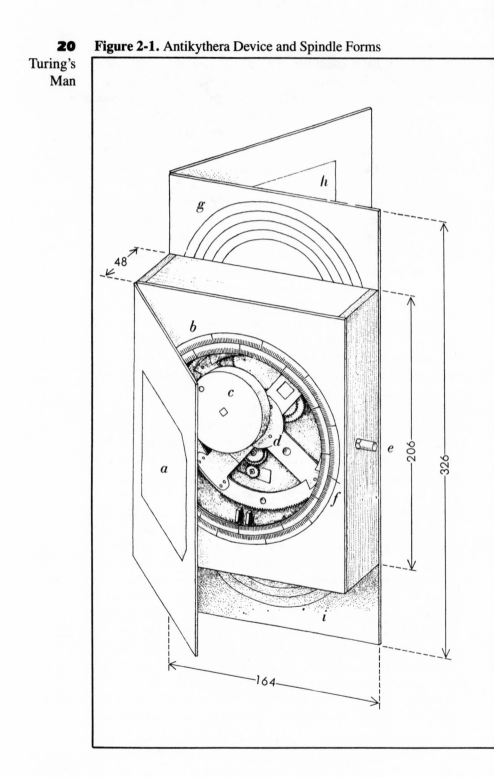

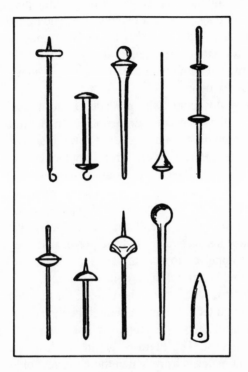

Left: The Antikythera device, found off the shore of a Greek island, was apparently a mechanism for calculating astronomical information. In this reconstruction by Derek J. de Solla Price, the numbers indicate dimensions in millimeters; the letters indicate parts of this complicated machine. The machine is an aberration; it seems out of place in the manual technology of the ancient world. This diagram is taken from "An Ancient Greek Computer," *Scientific American* (June 1959), 62. Copyright © 1959 by Scientific American, Inc. All rights reserved; reprinted by permission of *Scientific American.*

Right: The spindle was a defining technology of the ancient world. These spindle forms range from Neolithic to Roman times. The bulge in the shaft for stability is the *whorl*. The device could not be simpler, yet its symmetric motion as it twists yarn into thread is highly suggestive. For Plato the universe was a spinning shaft with eight concentric whorls. Diagram is taken from R. J. Forbes, *Studies in Ancient Technology* (Leiden: E. J. Brill, 1964–), 4:153; reprinted by permission of E. J. Brill.

mathematicians, yet without the help of such men, particularly the mathematicians, it could never develop such foundations. Although ancient "scientists" (philosophers, orators, historians) were not much interested in technology for its own sake, they did not escape its influence. Analogies from contemporary crafts were common in poetry and philosophy. This should not surprise us, for the spindle and the potter's wheel were as much a part of a philosopher's formative experience as the famed colors of the Aegean, the rugged mountains, or the inquisitiveness and garrulity of the Greek people.

The natural philosophers who preceded Plato in the sixth and fifth centuries—the Milesians, Empedocles, the atomists—had, as far as their thought can be recovered, little bias against technology, and their interests were firmly rooted in the concrete world, which they examined with a craftsman's combination of practical experience and poetic insight. Plato himself, even in the moral and political philosophy of the *Republic*, often invoked analogies from the crafts. For Plato, the craftsman was an individual who knew his particular field and performed one task well. Such expertise appealed to the philosopher's tidy sense of how a just society should operate: his utopia was a state composed entirely of craftsmen, with rulers who trained for that task all their lives—the original technocracy. Although he disliked manual labor, Plato was attracted by the craftsman's skilled creation from a preconceived pattern. The Platonic world of ideas was really a series of perfect patterns from which the imperfect objects of the material world were derived.

In the beautiful cosmological myth that ends the *Republic*, Plato described the heavens revolving about a "spindle of Necessity": an adamantine shaft and hook support eight whorls that are the celestial spheres. Here he gave a place of honor to the spinster Fates, who spin, measure, and at last cut the thread of life. They are among the most compelling figures of Greek mythology: "the daughters of Necessity, dressed in white, wearing garlands on their heads, Lachesis and Clotho and Atropos, sing in unison with the Sirens—Lachesis of things past, Clotho of the present, and Atropos of things to come" (*Republic*, 617C, my translation). The *Timaeus*, an influential, late dialogue, again invoked technology to explain the universal order, speaking of the creator of the cosmos as an ideal craftsman who formed the sphere of the fixed stars by turning it as a carpenter does wood on his lathe and mixing metaphysical elements as a potter mixes his clay. Man in the Timaeus myth was a miniature living world fashioned by lesser divine craftsmen: to make bone, they kneaded matter as a baker does bread and moistened it with marrow, and to form the skull, they turned bone on a lathe.

Plato's myth crystalized the meaning of manual technology for the ancient world. Aristotle and later writers were often more receptive to technology, but no one described better than Plato the meeting of divine form and terrestrial matter, the intelligent compromise between the ideal and the possible, that is characteristic of Greek craftsmanship. Aristotle expanded upon the analysis of reality as form and matter, and he was quite open to the analogy of the skilled craftsman. The form of a thing, the sum of qualities that identifies a table as a table, was for Aristotle imposed upon shapeless and indistinct underlying matter, just as the potter gave shape and color to indistinct clay. Aristotle's emphasis on the final cause or purpose for which a thing is created also reminds us of the craftsman's sense of purpose, the intelligence that formed the clay into a useful pot.

Let me return to the question of motion: here too manual technology influenced philosophy. Since nearly all motive power in the ancient world was supplied by human and animal muscle, it was natural for even sophisticated Greek philosophers to believe that what moved was alive. Plato wrote that "every body which receives motion from an outside source is lifeless, while every body that moves from within itself is alive, since motion is the very nature of the soul" (*Phaedrus*, 245E, my translation). His craftsman god in the *Timaeus* created a living (because moving) universe, whose stars and planets were also divine beings. Aristotle too identified movement as one of the characteristics of the soul: his god was pure thought or self-contemplation, which he regarded as the purest form of activity and the highest expression of life. Except for the Epicureans, who were heterodox in many respects, ancient philosophers were never far from animism. "For the early Greeks quite simply, and with some qualification for all Greeks whatever, nature was a vast living organism, consisting of a material body spread out in space and permeated by movements in time; the whole body was endowed with life, so that all its movements were vital movements; and all these movements were purposive, directed by intellect" (Collingwood, *The Idea of Nature*, 111).

The ancients saw their world as an expression on a cosmic scale of the principles of manual technology; their world was

formed by the intelligence of the universal craftsman but was not
mechanistic in the modern sense. The emphasis was not upon the
careful interaction of mechanical parts as in a clock but rather
upon the living, breathing whole.

Mechanical Technology and Western Europe

The disintegration of the western Roman Empire meant a drastic
change of outlook for the peoples living in its former provinces.
After a few centuries of retardation, medieval men began to assert their own technological identity; the stirrup, crop rotation,
and the moldboard plow soon made the medieval world different
from the ancient, socially as well as technologically. In matters
requiring central authority and resources, such as the building of
aqueducts, the Middle Ages could not equal the Roman achievement, yet in other ways did as well or better. Viewed with hindsight, the period seems to be one of transition in technology.
Many ancient skills survived, and new forms arose that would
culminate in the technology of the Industrial Revolution. Even by
the later Middle Ages, Western Europe was dedicated to the development of machinery and inanimate sources of power that
Greece and Rome had largely ignored.

The key invention was that of the weight-driven clock. The ancient world had contented itself with the sundial and the simple
water clock. Ancient mechanics such as Hero of Alexandria and
later the Arabs had described or built more complicated water-
driven mechanisms. There is a legend that Caliph Harun al Ra-
shid sent to Charlemagne in 807 a mechanism that not only told
the time but also worked a sequence of bells and moving horse-
men. Water clocks were less useful in northern climates because
they could easily freeze in the winter. It was clearly an improvement to devise a mechanism that depended only on gravity for its
operation. But since a weight on a rope falls with a constant acceleration, not a constant velocity, the weight-driven clock would
only work if this accelerating motion could be transformed into a
series of equal motions at regular intervals. This was the function
of the *verge escapement* with a *foliot* or vertical spindle, an ingenious mechanism typical of the artificial regulation of natural
forces that would characterize Western European technology (figure 2-2).

The first public clock that struck the hour was erected in Milan

Figure 2-2. Weight-driven Clock

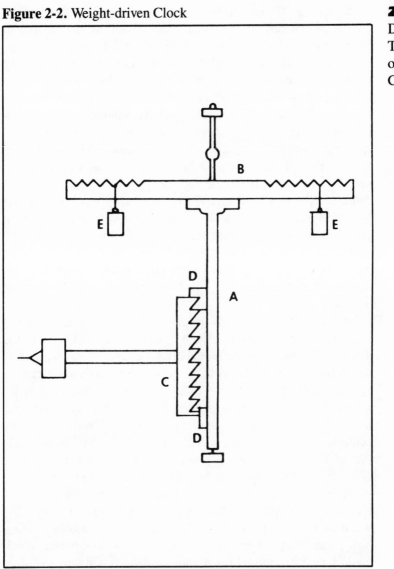

The weight-driven clock with the *verge* (A) and *foliot* (B) escapement was the defining technology of the mechanical age. As the verge or spindle swings back and forth, the pallets (D) engage the crown-wheel (C) one at a time. The crown-wheel turns at regular intervals because of an attached driving weight (not shown here) and so provides the regular movement needed for the rest of the clockwork. Even in its crudest form, the clock exemplifies the autonomous manipulation of the forces of nature that is practically absent from Greek technology. This diagram is taken from Samuel L. Macey, *Clocks and the Cosmos: Time in Western Life and Thought* (Hamden, Conn.: Archon Books, 1980), 20; reprinted by permission.

in about 1335. The elaborate clock at Strasbourg, built around the middle of that century, had a calendar as well as three animated Magi and a cock that crowed. Many others followed. The original appeal of the clock mechanism in the cities of Europe is only partly explained by convenience. Early clocks were inaccurate by as much as half an hour a day, and had to be reset frequently with the aid of a sundial; breakdowns were numerous. The energies of the early masters seem to have gone as much into ornamenting their clocks as into making them more precise, for the construction of human and animal automata, as well as astronomical indicators, was quite popular. Finally, in the seventeenth century, the Dutch scientist Christian Huygens created the pendulum-regulated clock that resulted in a vast improvement in precision. By then the clock was ingrained in the European way of thinking as the artificial model of the cyclical processes of nature, and it was ready to serve as the prototype for the mechanization of a variety of tasks that had previously been performed by hand.

Lewis Mumford reminds us that we cannot overestimate the importance of the weight-driven clock: "The clock . . . is the key-machine of the modern industrial age. For every phase of its development the clock is both the outstanding fact and the typical symbol of the machine: even today no other machine is so ubiquitous. Here, at the very beginning of modern technics, appeared prophetically the accurate automatic machine which, only after centuries of further effort, was also to prove the final consummation of this technics in every department of industrial activity" (*Technics and Civilization*, 14). This triumph of mechanization provided a powerful technological metaphor: the comparison of the world as a whole to the weight- or spring-driven clock. The value of the metaphor lay in the clock's cyclic motions, regularity, and independence from outside forces. The perfect clock could repeat its time-telling motions indefinitely, although in fact no real clock could approach this standard of perfection, at least until the invention of the pendulum regulator.

Clockmakers were quick to see the possibilities; as early as the fourteenth century, their devices were equipped with indicators for the position of the moon and the planets. In the Ptolemaic system of astronomy, the spheres of the planets, sun, moon, and stars circled the earth in eternally fixed courses. Their immutability impressed the ancients as much as it did medieval astronomers, who inherited the Ptolemaic theory. The theory itself had something of the complexity of a clock mechanism, for the spheres and motions multiplied as astronomers gathered more data. The account was never scientific in the modern sense: no attempt was made to explain the physical nature of these spheres in terms of terrestrial matter. In fact, it was central to the theory and to medieval physics in general that beyond the lunar sphere there were no ordinary substances—no earth, water, air, or fire, but only ether. Still, the cyclic gears of the clock made a fine analogy to the circling spheres of the heavens and made possible mechanical illustrations of the theory such as celestial clocks and armillary spheres. As early as the fourteenth century, Nicole Oresme saw that "the situation is much like that of a man making a clock and letting it run and continue its motion by itself. In this manner did God allow the heavens to be moved continually . . . according to the established order" (*Le Livre du ciel et du monde*, 289). Oresme was the first in a long series of thinkers to be inspired by the clock as a miniature universe.

What kind of a universe did the clock suggest? A precise and ordered cosmos, for the clockwork divided time into arbitrary, mathematical units. It encouraged men to abstract and quantify their experience of time, and it was this process of abstraction that led to the creation of modern astronomy and physics in later centuries. Copernicus's heliocentric theory, set forth in 1543, explained the courses of planets more simply, but not more accurately, than the Ptolemaic theory. In the early seventeenth century, Kepler applied to that theory a new passion for exact measurement and mathematical calculation. In the same period, Galileo formed his conviction that the book of nature was written in the language of mathematics, and by the end of the century, Newton had shown convincingly that terrestrial and celestial objects obeyed the same laws of dynamics, uniting all phenomena through his mathematical theory of gravitation. All these advances would be unthinkable in a culture without clocks.

The clock made explicit a view of the universe that orthodox Christianity had been tacitly encouraging for centuries. A. N. Whitehead pointed out long ago that the church actually encouraged the growth of modern science by insisting that the world was an ordered creation of God accessible to human reason: "Every detail [of nature] was supervised and ordered: the search into nature could only result in the vindication of the faith in rationality" (*Science and the Modern World*, 12). The church's rejection of ancient animism went hand in hand with this commitment to reason. God had created the world but remained separate from it, and this sense of separation encouraged men to study that world in a detached and later scientific way. Gone was Plato's divine, living universe, and in its place was a divine clockwork—Kepler himself had described the change in these terms.

In the seventeenth century, the clock metaphor was applied not only to the macrocosm of the world but also to the microcosm of animals and the bodies of men. It was the triumph of mechanism in its most radical form. The true mechanists, such as Descartes, Huygens, and Leibniz, explained all of nature (except for minds) in terms of interactions of matter. Descartes was certain that this explanation applied equally well to the growth and motion of plants and animals. In his *Principles of Philosophy*, he recognized no difference between the task of the clockmaker and that of the natural scientist, claiming that "it is no less natural for a clock, made of various gears, to indicate the hours than for a tree, sprung from a certain kind of seed, to produce a certain kind of fruit" (*Principia Philosophiae*, 326, my translation). Descartes's scientist was a detective who probed the unknown mechanisms of nature, including those of the human body, but not the workings of the mind, for here the mechanical analogy failed. It was precisely the fact that Descartes's nonmechanical mind drove a mechanical body that made it difficult to understand the relation between the two.

Leibniz had a more subtle explanation of that relation. For him minds were everywhere and behaved like incorporeal clocks. Like a vast timepiece, the material world has run on its predestined course since the beginning of time, and the mental clocks have also run on their course. The two sets of clocks have always been in a *"pre-established harmony*—pre-established, that is, by a Divine anticipatory artifice" (*Philosophical Writings of Leibniz*, 115). If a man cuts his finger, he feels pain, but not because the damage to his body has caused the pain in his mind. Rather, his mind, his mental clock, was originally constructed to feel pain at that precise instant, and his body and other bodies were arranged to cause him to cut his finger at the same instant. Although there is really no interaction between bodies and minds, there always appears to be perfect accord. At least one important thinker of this era, then, was so enamored of the clock metaphor that he was prepared to extend it even to the mind. But Leibniz's theory also reveals the major limitation of the metaphor when applied to human beings. All clockworks are fixed mechanisms; all their responses must be predetermined and built into the mechanism of gears and weights by the original maker. The course of the planets seems unchanging and free of interference (especially after the work of Laplace), so that it can easily be mimicked by a clock. But the human mind seems more changeable, seems capable of responding in a variety of ways to new circumstances. Leibniz's mental clocks were constructed (programmed) at the beginning of time with a set of responses to all the situations that they would ever confront. It was only because Leibniz was a determinist that he could program his mental clocks in advance.

Dynamic Technology and Western Europe

The clock is a beautiful, intricate, and suggestive device, but it can do nothing concrete to improve men's existence. It cannot lift water, mill grain, or produce textiles, although clockwork mechanisms may be used to regulate all of these processes. In Western Europe the technology of power developed in intimate contact with mechanical technology. Indeed, they were two manifestations of the same idea: the replacement of human craftsmanship with nonhuman devices. Mechanical technology is the artificial control of technical processes; dynamic technology is the harnessing of inanimate sources of power to drive the new mechanisms.

Dynamic technology also began in the Middle Ages with the exploitation of water and wind to do work. The Romans had known of the effective "Vitruvian wheel" as early as the first century B.C. and yet had ignored its implications. By contrast, the waterwheel attached to a mill was a common sight on the medieval manor. The power of falling water also turned saws for cutting wood, worked hammers for felting, and sometimes provided ventilation for mines. Where running water was not available, men harnessed the wind. The windmill was particularly important on the plains of England, the Low Countries, and northern Germany. With these prime movers and a variety of mechanical applications, the technology of the late Middle Ages and the Renaissance was considerably advanced and very different in character from that of ancient times. One great difference was that the moderns were constantly refining and redesigning their machines to be more efficient. The taste for technical innovation grew sharper than it had been in the ancient world, where men and ani- mals offered less opportunity for improvement and where im- provements even in the harnessing of animals were ignored.

The much later invention and perfection of the steam engine only made the new way of thinking more apparent. Atmospheric engines, which depend upon heat and pressure generated by gasses, were one of the first technologies for which the founda- tion had been laid by true scientific research. The work of physi- cists in the seventeenth century had established the existence of atmospheric pressure, and von Guericke's vacuum pump had clearly demonstrated that if atmospheric power could be properly harnessed, it would easily outstrip animal power. Thomas New- comen had built the first successful steam engine to drain coal mines as early as 1712, but the machine itself required so much coal that it was economical only where coal was abundant and cheap—in the mines themselves. James Watt's engine, patented in 1769 (with the key innovation of a separate condenser), first made economical use of the steam heated by coal. His machine achieved this result at the cost of being a complex mechanism, at least by contemporary standards. The piston had to be carefully bored, the valves allowing steam to enter and leave the piston re- quired careful adjustment, and the condenser itself had con- stantly to be pumped free of water. The machine was a precise and to some degree self-controlled device for transforming heat energy into useful power.

Dynamic technology, then, was as old as or older than mechan- ical technology, but it needed much longer to mature. Although accurate pendulum clocks were being built in the seventeenth century, it was not until the end of the eighteenth that the inani- mate prime mover found its place as a defining technology. The triumph of this technology was the steam engine. As a clockwork mechanism capable of producing power, it combined two quali- ties that had long before been expressed separately in the clock and the waterwheel. Although heat produced by coal was as much a natural resource as the flow of water in a river or the wind across the plains, the steam engine seemed by its very intricacy more artificial than the windmill or waterwheel, an expression of man's ingenuity and not a mere harnessing of nature. The engine could be used anywhere at any time; power no longer depended upon where nature had put a stream or when the wind was strong. It also encouraged a more mathematical view of natural sources of energy. John Smeaton's studies of the efficiency of waterwheels were made as the steam engine was being developed, and he soon applied his techniques to the new device. James Watt himself was led to define a horsepower in order to express the capacities of his engine.

The late eighteenth and the nineteenth centuries put the steam engine to a variety of uses in line with the spirit of emancipation from nature. The idea of synthesized power (along with mechanization) was applied in this era to crafts that had been plied entirely by hand in the ancient and medieval worlds. The change from the craft of making cloth to the modern textile industry (Arkwright's water frame, the spinning jenny, Cartwright's power loom, and so on) may have been the key to England's industrial revolution, but it also destroyed or at least distorted one of the richest sources of technological metaphor. The hardness and durability of iron and steel, available for the first time in large quantities, made possible the creation of machine tools, such as the precision lathe, that fashioned parts for other machines. The idea of machines that made machines introduced a new level of abstraction into technology. It was characteristic of the ancient craftsman to work close to his materials with a few simple tools. In the nineteenth century, a man would still use tools but would use them to build a lathe that would in turn cut screws to be used in a steam engine, the final product of his labors.

Meanwhile the philosophical appeal of technology continued, and the full impact of the Industrial Revolution with its dynamic technology made mechanisms of the previous centuries seem quaint. A new metaphor was added to that of the clock. The scientist and student of technology Norbert Wiener reminds us that in the nineteenth century the living organism was conceived of and studied as "a heat engine, burning glucose or glycogen or starch, fats, and proteins, into carbon dioxide, water and urea" (*Cybernetics*, 53). On the whole, however, dynamic technology proved more compelling as an analogy for macrocosmic events than for the microcosm of life. The steam engine was an artifact for transforming heat into mechanical energy, but it relied upon universal principles of thermodynamics, gradually explained by Carnot, Clausius, and Kelvin. Writing in 1824, Sadi Carnot was enthusiastic in drawing analogies between natural "heat engines" and synthetic ones: "It is to heat that we must attribute the great and striking movements on the earth. It causes atmospheric turbulence, the rise of clouds, rain and other forms of precipitation, the great oceanic currents . . . lastly it causes earthquakes and volcanic eruptions. From an immense natural reservoir we can draw the motive power we need. . . . To develop that power, to appropriate it to our own use is the purpose of fire-engines" (from *Reflections on the Motive Power of Fire*, quoted by Cardwell, *Turning Points*, 129). D. S. L. Cardwell says of this passage from Carnot: "If, to the seventeenth century philosophers the universe seemed like a gigantic piece of clockwork, to nineteenth century thinkers it was to appear to have many of the attributes of a heat-engine" (*Turning Points*, 130). Those attributes were not always encouraging ones. The second law of thermodynamics suggested that the universe was an imperfect steam engine, gradually running down, dissipating its organization and giving up its energy as useless heat. The brutality of this new metaphor reflected the brutalizing conditions to which the Industrial Revolution subjected the working class in Europe and North America. Contemporary astronomers could chart the course of the planets with far greater precision than Newton had done, but the age seemed more obsessed with the dynamic than the mechanical aspect of the European technological achievement, that is, with the pursuit of power. The huge steam engines at the end of the century, which produced thousands of horsepower, were controlled by complex arrangements of gears, but their purpose was to pull rail cars and to drive ships. By comparison, a clock must have seemed a benign and rather idle device.

Electronic Technology

The antecedents of computer technology date back at least as far as the seventeenth century, when mechanisms were already being built to perform arithmetic. Pascal designed an adding machine, Leibniz a wheel to perform the four arithmetic operations. A glorious but doomed attempt at a clockwork computer was made by Charles Babbage, a British mathematician of the nineteenth century. He first planned a Difference Engine, a gear-driven calculator based on Newton's method for doing mathematical integration; another engineer in fact built a working version. Babbage himself went on to dream of something more grand: the Analytical Engine, a mechanical computer that could be programmed for all sorts of mathematical problems. Only parts of the machine were ever constructed since machine tooling of the time could not manage the needed precision. In general, Babbage's design was too complex; he sought to create a logic machine of high precision without proceeding through a series of prototypes as a modern engineering project would.

Nonetheless, Babbage and his protégés, among them the Countess of Lovelace, Byron's daughter, were genuine visionaries. In their writings we often find expressions of a world view fully a century ahead of its time. If the Analytical Engine had been built, it would indeed have been the first computer, but Babbage was trying to fashion out of clockwork a device that really belongs to the age of electronics. His refusal to accept the practical limitations of his materials and his century may be typical of a visionary. (He once said he would give up the rest of his life for the privilege of spending three days in the age five hundred years hence.) However, the whole point of a logic machine is that it must function; it must embody logic in an assembly of physical components that work together harmoniously. As we shall see repeatedly, a computer makes new notions of logic, time, space, and language accessible and impressive to a vast new audience precisely because it embodies these ideas in a machine. A computer that does not work is remarkably unimpressive (as any programmer knows whose machine has "gone down" during a demonstration before a skeptical audience). Babbage's blueprint and disassembled parts could not change the world. His writings give proof of this sad fact: they speak eloquently of instruction steps, programming logic, symbol manipulation, the limits of machine time. Yet the scientists of the age apparently did not feel the significance of the message. Babbage remained a brilliant aberration, a prophet of the electronic age in the heyday of the steam engine.

The digital computer is the great achievement of twentieth-century technology. Before and during the Second World War, a number of engineers in America and some in Europe were working on electromechanical calculators and planning for fully electronic ones. After the war, electronic components made from vacuum tubes reached a new degree of refinement, and the transistor was invented in 1949. These devices were made to operate reliably at speeds that had never before been approached in the world of engineering, speeds previously belonging exclusively to the physicist. They also possessed a remarkable versatility: each could change its function as quickly as voltage could be changed within a circuit. Machines could now be built to perform arithmetic and logical calculations; these would not be like the old mechanical calculators in immediate contact with a human operator but would work from a set of previously coded instructions. Even early computers were extremely intricate in design, thousands of times more complicated than the most carefully crafted steam turbine. Unlike previous machines, they could manipulate their very internal structure, the patterns of electrons in their circuits, to express mathematical and logical relationships. In theory, one can make a digital computer out of metal gears, as Babbage planned to do. In practice, clockwork is too cumbersome, breaks down too often, and moves too slowly to build a computer that is both capacious and efficient. Twentieth-century mathematicians such as Turing and John von Neumann realized that they could create from electronic components what Babbage had failed to make from gears.

Von Neumann helped to draw the blueprint for the digital computer, and his name is still used for the design that most digital machines follow. The von Neumann machine is made up of a *central processing unit*, where arithmetic and logical operations are performed, and a storage unit (*memory*), where instructions and data are kept awaiting their turn in the processor. Storage and processing are fully electronic. The memory consists of some electronic medium (now generally transistors) that preserves the data in its *binary code* (as a series of 0s and 1s), and the machine operates by shuttling electronic information (at speeds approaching that of light) back and forth between the memory and the central processor. The operation is intricate; digital computers are the most complicated devices ever built by men. Biological structures are more highly organized because they begin at the molecular level with patterns of amino acids. Still, the computer is one of the first human artifacts with parts too small to be seen with the unaided eye.

The computer itself is only one of a family of electronic devices: recording equipment, measuring instruments of all sorts (for physics, chemistry, and medicine), radio, and of course television. These devices in turn stand at the top of the vast technological pyramid of our society. They depend for their existence and continued improvement upon the mass production economies of Europe, Japan, and North America, and they have acquired some of their characteristics from this dependency. The von Neumann machine is the ultimate assembly line, manipulating parcels of information at electronic speeds. Computers themselves are made by an industry demanding specialized labor and expen-

Figure 2-3. von Neumann Schematic

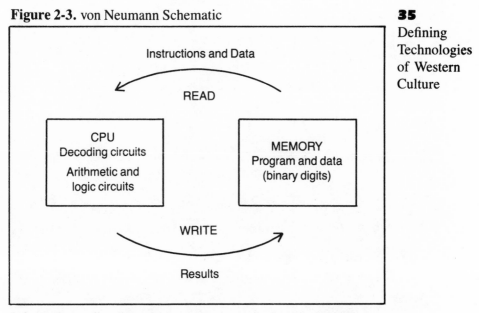

In its roughest outline, the von Neumann computer simply reads coded information into the central processing unit (CPU) and writes information back after computation. All information (instructions, input data, and results) are expressed as sequences of binary digits.

sive machinery. Integrated circuits are constructed in laboratories whose air is filtered thousands of times cleaner than the air in hospital operating rooms, yet something like half of the circuit chips made must be thrown away because of defects. It is no accident that electronic technology flourishes only in those countries with diversified and industrialized economies.

To produce and exploit a device as highly organized as a computer itself requires a high degree of organization: many people must work together because no one can master all the separate areas of knowledge. Computer technology is team technology. It was born through collective projects at such universities as Pennsylvania, Cambridge, and Harvard, where men and women with varied backgrounds in science and engineering were united by the common vision of creating a logic machine. Since then, every new machine and most significant programs have also been products of a team effort. In this respect, too, the computer is an archetype for current technology: the twentieth century has been shaped for good or ill by great team projects in engineering, of which the three outstanding examples are the V2 project in Germany, the Manhattan project in America, and more recently the

Figure 2-4. Microcomputer

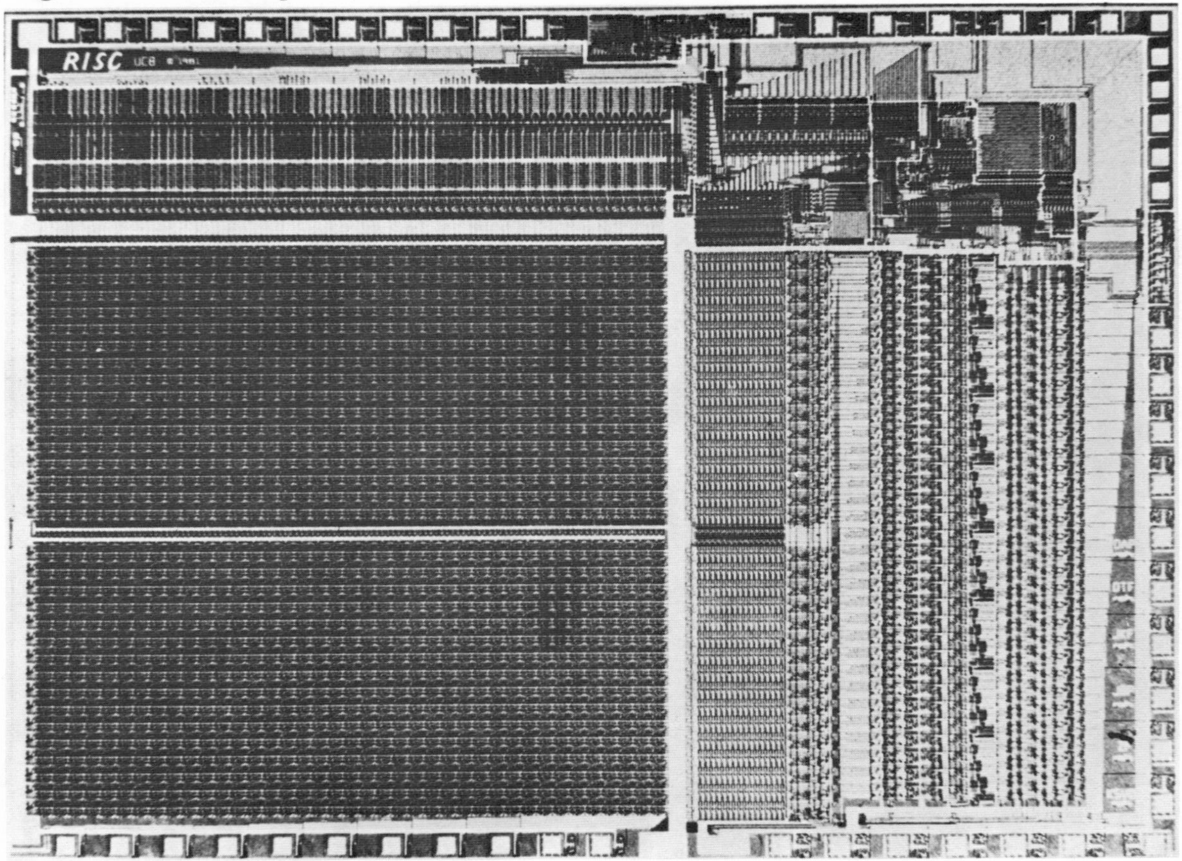

The microcomputer on a chip is a marriage of geometric form with the principles of electronics; there is more than a little beauty in its pattern of parallel and intersecting lines that store, transmit, and operate upon electronic information. This particular machine is called "RISC" and was built by David A. Patterson and his students in the Department of Electrical Engineering and Computer Sciences at the University of California, Berkeley. It is an especially good example of regularity and simplicity of design. Photo reprinted with the permission of David A. Patterson.

American space program. Even as late as the nineteenth century, individual inventors could make important advances working alone. Today the man or woman who gets the credit for an advance is often simply the group's leader.

The computer is a product of the belated but fruitful marriage of science and technology. The machine is built upon the work of the best physicists of the past two centuries, although it depends as well upon the empirical sense of generations of electrical engineers. Ever since the seventeenth century, science and practical experience have been combined to advance technology, but the link was never so close and the debt to theoretical science never so great as in electronics since the Second World War. This field, along with aerospace technology, is the culmination of the rapprochement that began in the age of Descartes and Newton. Today science and engineering are often indistinguishable. James Watt needed a certain amount of physics to improve the Newcomen engine. Steelmakers in the nineteenth century profited from the advice of professional chemists. But anyone today who wants to design a new transistor does not go to a physicist for advice; he is a trained physicist in the first place. In this way too the computer is an archetype, for everywhere in the world of high technology we find the influence of advanced physics, chemistry, or mathematics.

From the Clock to the Computer

The ancient and Western European civilizations each possessed a characteristic set of materials, techniques, and devices, which not only provided the physical means of life but also helped to shape their cultural outlook. It is against this background—the manual technology of the ancients and the mechanical and dynamic technology of the Europeans—that the computer's impact can best be evaluated. The computer shares qualities with many of its technological predecessors, and yet its combination of qualities is unique.

To begin with, if we think of a technology as the controlled application of power to manipulate the environment, it is not immediately clear that the computer is a technological breakthrough at all. The computer does no work in the nineteenth-century sense, makes no apparent change in the physical environment. It can serve as a regulator for more conventional machines, but in any case it is the information a computer provides that makes it useful. In that respect the machine is an extension of the mechanical rather than the dynamic aspect of Western European technology, of the clock rather than the steam engine. The clock provided information more conveniently and eventually more accurately than the sundial had done before, but it did not intrude physically upon life except by ringing the hour. The usefulness of its information was perhaps not obvious; improved information is seldom as obviously useful as tangible improvements such as a better plow or a more efficient loom. Like early computers, early clocks were expensive and delicate, and it required a major corporate effort to acquire one. As clocks became common, however, they became not merely useful but unavoidable. Men and women began to work, eat, and sleep by the clock, and as soon as they decided to regulate their actions by this arbitrary measurer of time, the clock was transformed from an expression of civic pride into a necessity of urban life. The computer too has changed from a luxury to a necessity for modern business and government.

As an information processor, the von Neumann computer goes far beyond the clock; it is at once mechanical and nonmechanical. Like the weight-driven clock, it obeys prescribed rules and performs its operations in a fully determined sequence. In fact, the central processor of the computer contains within it an electronic clock, whose extremely rapid pulses determine when one operation has ended and another is to begin. Still, the processor has no moving parts: it achieves its calculations by manipulating electric currents, not by a subtle interplay of gears and levers. Although the immediate forerunners of the postwar computers were huge calculators made of telephone relays, all modern electronic computers operate, as the name suggests, upon electrons. They can be made fast and compact precisely because their components are so small and fast moving. Central processors are smoothly running, silent, motionless machines, whose calculations the programmer can neither see nor feel directly.

Electricity and magnetism, which act on bodies at a distance, lend a sense of mystery to any device that they power. "For a long time," writes the historian of technology Siegfried Giedion, "a tinge of the wondrous seemed to pervade all things electric. There was in truth something to marvel at when the seventy-year-old Michael Faraday toured the English lighthouses in 1862 and first beheld the practical application of his light, or 'magnetic spark' as he called it, which had arisen in his hands three decades

earlier" (*Mechanization Takes Command*, 542). How much more
astonishing would Faraday find the modern calculating engine, solving differential equations with his magnetic spark? Altogether, this mysterious action-at-a-distance seems more at home in the ancient world of animism than in the materialistic nineteenth century. In fact, with its clean lines and pure geometric quality, the computer reminds us more of a Greek vase or chair than of a noisy steam engine. In both blueprint and physical embodiment, the computer is almost pure design, its rows of circuit boards printed with metallic lines crossing and recrossing in rhythmic patterns. However symmetric and tidy a steam engine or even a clock may be in blueprint, the actual result is oily, noisy, or inexact, simply by virtue of the materials used.

Unlike the clock or the steam engine, the computer is not a fixed mechanism. The genius of the von Neumann machine is that the program (operating instructions) and the data are stored in the same binary code and loaded together into the memory, and this coding means that the program can be altered as easily as the data, indeed, that there is no logical difference between the two. If we made an analogy to the weight-driven clock, the data would be the force that the weight exerted on the escapement wheel. This force was translated by means of the verge and foliot mechanism (and by other gears) into the movement of the hands of the clock. The gears, then, were the program—their single purpose was to produce as constant output the correct time. Any change in the program required the clockmaker to readjust these gears. In the steam engine, the coal used to fire the boiler and perhaps the steam itself were the data. All the mechanisms of the boiler, the piston, the condenser, and the rest were the program, whose purpose was to translate this input energy into a mechanical output. The steam engine could adjust itself to a certain extent during its operations: one of Watt's inventions was the "centrifugal governor," which regulated the flow of steam to the engine depending upon how fast the rotary axle was turning. Any more drastic change in the machine had to be made by laborious human intervention.

By contrast, the computer can be changed simply by loading a new program, and each new program makes the computer into a new machine. A programmer is a designer who has the remarkable advantage of being able to test his design as soon as it is specified. For the design *is* the program, written in a suitable language such as PASCAL, and he need only submit the program to the computer to find his machine realized. Furthermore, since a program itself is simply a series of binary digits, the computer can be programmed to write its own programs. The equivalent process in a steam engine would be to throw the gears into the furnace along with the coal and to expect the engine to produce by itself a design for a new machine.

The Electronic Brain

The computer is succeeding the clock and the steam engine as the defining technology and principal technological metaphor of our time, chiefly because it can reflect the versatility of the human mind as no previous mechanism could do. Descartes was willing to compare animals to clocks because he believed that animals and such lower orders of life as trees were capable only of fixed responses (however subtle) to their environment. The human mind seemed to him to belong to another realm altogether. "While reason is a universal instrument which can serve for all contingencies," he tells us in the *Discourse on Method*, "these [mechanically created] organs have need of some special adaptation for every particular action. From this it follows that it is morally impossible that there should be sufficient diversity in any machine to allow it to act in all the events of life in the same way as our reason causes us to act" (*Philosophical Works*, 1:116).

Much of the feeling against extending the clock metaphor to the mind must have come from the conviction that this divine aspect of man could not be material, be made of interacting parts, or be extended in space or subject to decay. The Cartesians wanted to keep divine and human minds wholly separate from the material universe, and it is striking that even the followers of Newton, who believed that God was intimately involved in the world, eventually identified him not with matter and the clockwork system of the planets and stars but with pure, empty space. For them, too, simplicity seemed more divine than mechanical complexity.

Whatever their metaphysical reasons for disliking mechanical explanations of the mind, the natural philosophers of this era were not confronted with machines that reflected the human ability to respond to the world. The computer is now capable of a vast range of responses that mimic, although they do not match, human abilities. In the late eighteenth century, a man named von

Kempelen built as a hoax a chess-playing automaton, which concealed a man inside and astonished audiences with its performance. Today a chess-playing machine is no hoax, and, although it cannot yet beat a master, it does play winning chess against most amateurs. Along with this startling ability to react to changed circumstances, the computer has the potential to "learn," to change its program as it operates. The potential has never been exploited as some computer specialists have hoped, that is, to create a simple computer-child and then to teach it much or all adult human knowledge. But even if this hope belongs to the realm of science fiction, it still serves to enhance the identification of the computer with the human mind.

The electronic digital computer has overcome in part the objections to the materialization of the mind. The analogy is all the more compelling because the brain itself is known to consist of electrically sensitive cells. After the potter's wheel, the clock, and the steam engine, the computer marks in one way a culmination of the trend away from animism. The highly animistic Greek view was that life existed everywhere in the universe where natural motion could be found, a view that allowed for perfect unity between man and the cosmos. The craftsman metaphor illustrated both: deities like potters fashioned men as they fashioned the planets and suffused both with life. The orthodox Christian of the Middle Ages reacted against animism. His God had chosen to remain separate from the world, and so the mechanical clock seemed an appropriate model for the planets and even for animals, which, though alive in a sense, had no share in divinity. Only the human mind or soul was excepted, for it remained linked to God. But once Descartes and Leibniz had eliminated God from the physical universe, he threatened to disappear altogether. Success in the physics, chemistry, and biology of this world made questions of theology and metaphysics less interesting or less important. Newton's planetary system seemed to need God to keep its orbits from perturbing one another, but Laplace demonstrated a century later that such perturbations had no long-term effects. He told Napoleon that in his astronomy there was no need of God as a hypothesis. With the computer, another step has been taken in this evolution of ideas, for we now have an inanimate metaphor for the human mind as compelling as the clock was for the planets.

The rapprochement has come from both directions. The computer is in some sense the most animistic, least mechanical of all machines. It operates with that mysterious source of power, electricity, and has a cleanly geometric structure without moving parts (at least in the central processor). On the other hand, the mechanics of the human body—for example, the way food energy is converted into movement—are much better understood than they were by the ancients. There has never been more evidence for regarding a human being as a vastly complicated machine than there is today, although the controlling agent, the central nervous system, remains in many respects a mystery. It is becoming more and more attractive to equate the mind and the brain and to identify both with the computer. Some biologists and psychologists have made the leap, using the language of electronic processors to characterize their fields of study. Many have so thoroughly absorbed the technology that they no longer claim to be explaining the brain in terms provided by the computer; instead, they say that human brains and computers are two examples of "thinking systems."

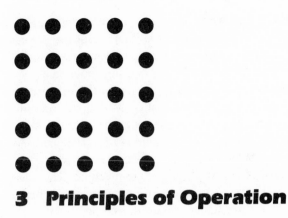

3 Principles of Operation

Every new computer is accompanied by a manual entitled something like "Operation and Reference." This explains the workings of the machine, usually not at the level of its electronics, but rather in terms of its logical parts and their function. The reader may care to think of the following chapters as a kind of cultural operation and reference manual for the computer. Each concentrates on one element of the technological world of computers—mathematics, logic, space, time, and so on—and shows how this element contributes to the making of Turing's man. Any good manual begins with an overview of the whole machine, and that is the purpose of this chapter: to set out the parts and function of the Turing and von Neumann machines for readers with little previous programming experience. (Others may prefer to go directly to the discussion of mathematics by computer.)

The Turing Machine: States and Symbols

A Turing machine is an abstraction, a creation of logic and mathematics. But it may also be thought of as a game, one that can be played with no more advanced technology than pencil and paper. To play any game, we agree to something like that suspension of disbelief that makes possible theater and the movies. We enter another world, a world whose logic consists entirely of the rules of the game. Insofar as we take the game "seriously," we concentrate upon the logical corners and alleys that the rules define, and we do not bother about experiences that fall outside of the game. In Monopoly, for example, players buy and sell property, collect rent, and move their pieces. The rules say nothing about their getting hungry, falling asleep, or suffering a myocardial infarction; dying is not part of the game. What matters is that buying and selling property, collecting rent, and moving pieces make up a logical and satisfying whole. A good game is self-contained. A Turing machine is just such a logically self-contained world.

The sole inhabitant of this world is a logic machine. The machine consists of two parts: a finite set of operating rules, built somehow into the works and unchangeable during operation, and a tape of unlimited length, upon which changeable information can be stored. In figure 3-1, the tape is divided into cells, each of which may contain only one symbol, and there is a marker to indicate which cell is being inspected at any given moment. The tape holds the information that the machine is to process; in our example it is binary information, for the three allowed symbols are "0", "1", and a blank. If the machine chooses to write any output, it must do so somewhere on the same tape. It may write over input information, and the number of cells available to the right is without limit. Now, the set of operating rules controls the action of the marker along the tape. In figure 3-1 these rules are written in English, although Turing used a logical notation to achieve the same meaning. Our example has only two rules; usually there are more, perhaps a dozen or a hundred, but the number must always be finite.

The machine is sequential; it operates by activating one rule at a time, performing the actions prescribed by that rule, and then activating another. Before and after each operation, the machine is said to be in one of a limited number of *states*. These states are defined by the logician who has constructed the machine; in our example they are named Q1 and Q2. The machine always "knows" two things: its current state and the current cell pointed to on its tape. These two factors are all that are needed for its operation, for the machine always chooses its next move by finding the rule corresponding to its current state and the symbol in the current cell. State and symbol make the machine go.

An activated rule calls for the following: writing a symbol on the cell designated by the marker, moving the marker one cell right or left, and changing the current state. These are the only means by which a Turing machine can manipulate data. The pro-

cess continues in this fashion until the machine activates a rule
that tells it to halt; it has then completed its task.

A Turing machine is a game in which no room is left for player initiative. The only element that is allowed to vary is the input tape. Once an input has been chosen, the machine evokes its rules with the regularity of clockwork: for each input there is only one possible outcome. In our example (figure 3-1), the tape is set with the input "1" followed by blanks, and the machine is started in state Q1. The first rule is applied, the machine writes the symbol "1" over the current "1" (in effect leaving the symbol unchanged), moves its marker one cell to the right, and enters state Q2. The current symbol is now blank and the state is Q2, so that rule two applies. The machine writes "0", moves its marker to the right, and halts. The program has ended; its modest goal was to replace the number 1 with the number 10.

There are of course an infinite number of other, more interesting Turing machines; the principles of operation are the same no matter what the machine is designed to do. The tape may hold the digits of a number, and the machine may embody rules for finding the square root of this number. There might be a series of numbers on the tape, and the machine might be designed to compute the thrust needed to put a spacecraft in lunar orbit. The really challenging game is to design the Turing machine in the first place, to come up with operating rules and states that will make the machine act as desired. It turns out that the logician does not have to construct a new machine, embodying new rules, for each new programming task. One machine, a universal Turing machine, can be designed so that, given the proper input symbols on its tape, it can perform any task that any individual Turing machine can perform. A digital computer can be thought of as a universal Turing machine.

Turing himself invented this game in 1936 in order to prove some abstruse results in symbolic logic. But his invention captured perfectly the theoretical structure of all digital computers to the present day. The word "machine" was used by Turing himself, although it traditionally meant a device for producing power to do work. A Turing machine does not perform work as a steam engine does; it merely moves its marker back and forth along its tape—examining, erasing, and writing symbols as it applies its rules of operation. Such a machine does not improve man's ability physically to manipulate his environment, and it is only by analogy that we should use the word "machine" at all. By any

Figure 3-1. Turing Machine

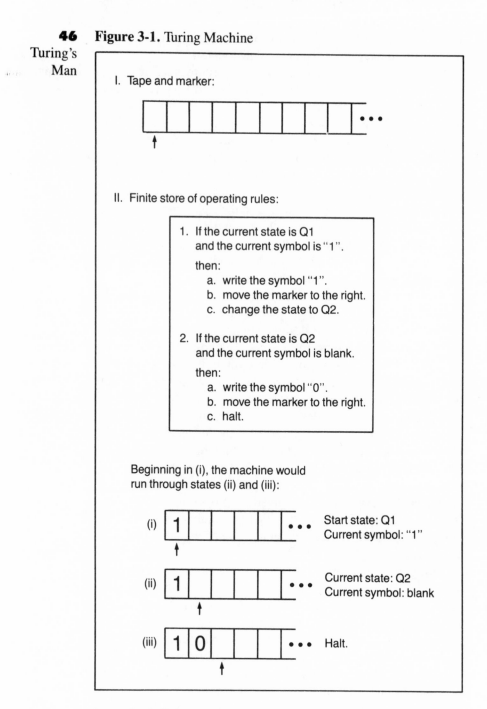

I. Tape and marker:

II. Finite store of operating rules:

1. If the current state is Q1
   and the current symbol is "1".

   then:
   a. write the symbol "1".
   b. move the marker to the right.
   c. change the state to Q2.

2. If the current state is Q2
   and the current symbol is blank.

   then:
   a. write the symbol "0".
   b. move the marker to the right.
   c. halt.

Beginning in (i), the machine would
run through states (ii) and (iii):

(i) | 1 | ••• Start state: Q1
Current symbol: "1"

(ii) | 1 | ••• Current state: Q2
Current symbol: blank

(iii) | 1 | 0 | ••• Halt.

A very simple Turing machine. Its task is to replace the symbol "1" with the symbol "10". In this, as in any Turing machine, each action is predetermined: given the current symbol and state, only one result is possible.

name, Turing's abstraction in fact shows what a computer does:
to "process information" by computer is nothing more than *to
replace discrete symbols one at a time according to a finite set of
rules*. Any task a digital computer performs (from figuring taxes
to playing chess) can be explained by this simple replacement
procedure, and any task that cannot be carried out under this procedure cannot be programmed in a digital computer. It can be
shown—and this was Turing's original point—that there are
problems no Turing machine can ever guarantee to answer; the
machine may go on infinitely spinning its tape without coming to
a result. Such problems pose an ultimate, though by no means the
only, limitation on the power of the digital computer.

The von Neumann Computer

We have traced the logical shadow of the computer in Turing's
machine. No matter how ingenious, that device was a game for
logicians and could only have had a limited impact on our culture
(like the other concerns of twentieth-century logic) if it had not
found a physical embodiment as the von Neumann computer.
About ten years after Turing first made his suggestion, John von
Neumann and his many colleagues realized its significance and
worked to apply Turing's logical scheme to the physical materials
from which information processors might be built. They sought
to make a Turing machine out of vacuum tubes instead of pencil
and paper; the electronic version would operate millions of times
faster than a mathematician applying his rules and erasing his
symbols accordingly.

Again, two parts needed to be considered: the rules of operation and the data upon which to operate. The data could sensibly
be put on punched cards, represented inside the machines by
some sort of electronic storage, and eventually dumped as output
onto punched cards or paper. The problem was with the rules of
operation, for the tendency had been to think of these rules as
fixed, to express them in the very structure, the wiring of the machine. Early machines offered plugboards, something like telephone switchboards, in which wires had to be reconfigured for
each new problem to be solved. The wires led out to various parts
of the machine, controlling the calculations that the data would
undergo. The operator literally had to wire a new machine, designed exclusively to execute that one problem.

The unique feature of the von Neumann computer is that the programs and data are stored in the same way, as strings of binary digits. The program, a set of instructions to add, subtract, or manipulate pieces of data, is loaded along with the data into the computer's memory. The *central processing unit (CPU)* of the machine distinguishes between instructions and data usually only by position. It decodes and executes instructions and operates accordingly upon the data. The illustration of the von Neumann computer in figure 3-2 is still quite abstract: in a current machine, each of the boxes would be electronic devices made of thousands or millions of transistors; the arrows would be wires or connections allowing electrical "information" to pass between the devices in the directions indicated. The box marked *memory* corresponds to the tape of the Turing machine: here the input is stored and the intermediate results are written by the CPU. The CPU corresponds to the operating rules of a universal Turing machine. Built into its electronic structure are rules for making sense of the instructions and data that it finds in the memory. The CPU goes through the memory, much as a Turing machine moves along its tape, picking out instructions one at a time and executing them.

The CPU is made up of logic circuitry and various registers. The registers hold individual instructions or small amounts of data for immediate processing. The control unit, more of a functional than a physical division in modern computers, breaks the instructions into a series of electronic actions, directing the step-by-step operation of the whole system. (This unit replaces the plugboard of the earlier machines.) The *arithmetic and logical unit (ALU)* performs the actual additions, subtractions, and logical business required.

Suppose the next instruction in the memory is to add two numbers and store the result somewhere else in memory. The sequence is:

1. The control unit signals the memory to send the next instruction and an instruction register to receive it. Decoding circuits read the instruction and discover that it demands an addition.
2. The control pulls the two numbers to be added from the memory and puts them into data registers.
3. It sends these two numbers through the arithmetic unit, where they are summed, and puts the sum back into a data register.
4. It sends the sum from the register back into memory.

The process is then repeated on the next instruction and so on through the program. The computer works in cycles, each of which has two parts: fetch and execute. The fetch step is the "reading" and interpreting of the instruction (1 in the above sequence), where the decoder discovers what is to be done. The actual processing (2 through 4) of the data is the execute step.

The cycle of fetch-and-execute proceeds with numbing regularity perhaps millions of times each second; it is repeated for each new instruction and new packet of data. For all its extraordinary speed, the von Neumann computer operates rather like a two-stroke gasoline engine, drawing its instructions in with the first stroke and executing them with the second. A closer analogy, perhaps, is an industrial assembly line or job shop, in which an endless stream of identical items are processed. Here the items are not automobiles or alarm clocks but electronically coded packets of information. These move at speeds that are healthy fractions of the speed of light back and forth between memory and processor; within the processor they are broken into smaller packets and recombined in electronic forms of arithmetic and logical calculus. In this fashion, the computer processes information: it begins with one string of binary digits (the program and the data) and ends by assembling another (the output) from fragments of the first.

It is the logical unity of the von Neumann, stored-program computer that makes it so powerful. New programs, new sets of instructions, can be loaded into these machines as easily as new data. Each program in effect makes the computer into a different machine, one with a new purpose, without any change in the wiring. The same physical equipment may serve first to calculate the orbit of a spacecraft, then to alphabetize a list of names, then to determine averages and deviations of a statistical sample. Since each of these tasks calls for a logically different Turing machine, the physical equipment that can accomplish them all is a universal Turing machine. Thus logic and electronics meet at precisely this point: the von Neumann computer.

Hardware and Software

The von Neumann design has encouraged the division of the electronic world into two groups: the electrical engineers and the logicians or programmers. The computer itself (transistors and

Figure 3-2. von Neumann Computer

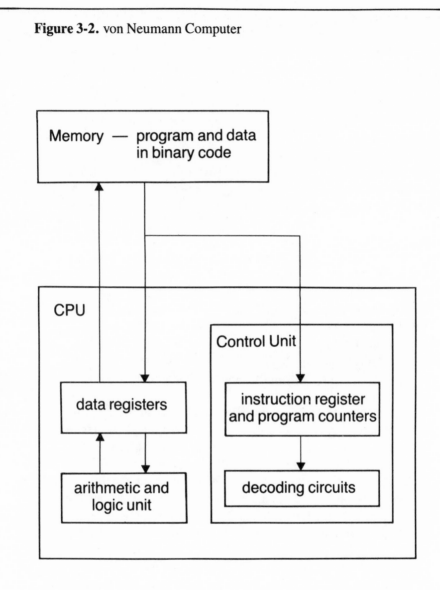

A skeletal diagram of the von Neumann computer: boxes indicate functional units and arrows indicate the flow of data among units. (The control lines, which carry signals from the decoder to activate other units, are not shown.) The computer's operation may be broken into the following steps:

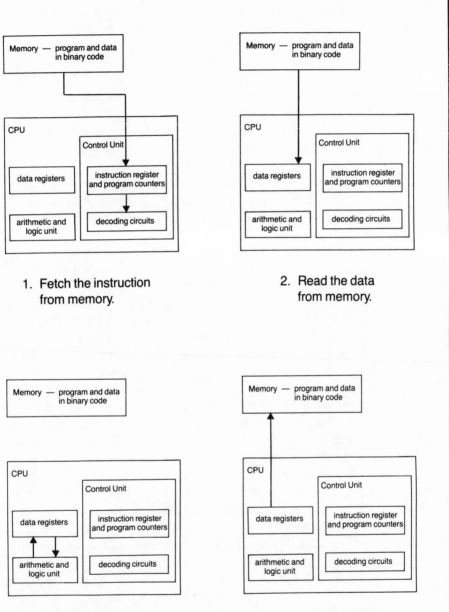

1. Fetch the instruction
from memory.

2. Read the data
from memory.

3. Execute the instruction
upon the data.

4. Return the result
to memory.

wires) and all the peripheral recording and printing devices (tape and disk drives, television screens, typewriter terminals) are called *hardware*, presumably because of the materials from which they are made and because they are components of the computer system that are "hard" to change. Replacing one tape drive with a faster one means at the very least rearranging a number of wires and pushing two heavy devices around the floor. Changing a feature in the central processor may require rebuilding the machine. Hardware is designed by engineers, though in collaboration with programmers who will put the computer to work.

The *software* is the whole set of programs that brings the hardware to life. Software is "soft" precisely because it can be easily changed. The hardware of a system generally remains the same, but the computer is always prepared to accept coded versions of new programming ideas. Each program is a recipe for computation, a sequence of instructions suitable for execution by a digital computer. The word is sometimes used synonymously with the word *algorithm*. More precisely, an algorithm is the logical idea behind the program, a strategy for solving a problem. This strategy may be represented in English, in mathematical symbols, or in familiar programming languages such as FORTRAN or PASCAL. The exact representation is not important, as long as the strategy is a procedure that a computer can follow. "Tell me something useful about the numbers 6 and 8" is not algorithmic. "Add 6 and 8; divide the sum by 2; and report the quotient" is an algorithm because it can be translated into a suitable programming language and run through the machine.

Are there problems that cannot succumb to the procedural techniques of Turing and von Neumann? What are the limits to the logic of state and symbol, fetch and execute? Turing's man is inclined to include a great deal of human intellectual activity under the rubric of programming, inclined to feel that most intellectual issues confronted by humans will eventually be computable. If Turing is right about artificial intelligence, then in the year 2000 "Tell me something useful about the numbers 6 and 8" will be an algorithm. But let us begin to examine what the computer can already do, first in the field of mathematics.

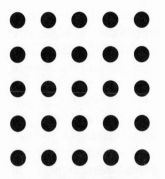

4 Embodied Symbol: Mathematics by Computer

In the twentieth century, the drive for speed and efficiency took hold of mathematics as it did so many other facets of our culture. The immediate ancestors of the digital computer were electromechanical calculators, which were themselves much faster than humans. The desire of scientists and the military for even faster computation led to the first fully electronic machines. The first project for the electronic computer ENIAC was to speed mathematical work at Los Alamos in connection with the atomic bomb. And scientists today are still seeking more computing power for ever more ambitious calculating tasks. Their demands spur designers on constantly to improve the "number-crunchers" or "supercomputers," as the most advanced and powerful machines are called. Some of the computer's most impressive achievements have come through number-crunching—reckoning in a few minutes or hours what would have taken human scientists centuries to calculate by hand. This was, after all, what Charles Babbage originally envisioned as the purpose of his Analytical Engine. Simple counting and higher mathematics are perhaps chief among the things that computers do well.

Yet computers have not simply taken over tasks once performed painstakingly by humans; they have also given a new significance to numbers. The computer allows, indeed seems to demand, quantification of all kinds of data, but the very concept of quantity with which it operates is different from the one that has prevailed at least since the time of Descartes. The difference stems from the fact that the computer embodies numbers and arithmetic operations in electronic circuits. No matter how fast or advanced the computer, its mathematics has a lack of precision, a quality of finiteness, which betrays its origin as a machine and radically separates it from the elegance of earlier Western mathematics. Computer mathematics is providing powerful new symbols for our age, symbols as important to the making of Turing's man as mathematics itself is to the making of electronic technology. Fortunately, we do not need to be trained mathematicians to understand the difference the computer makes, but we do need to look at how numbers are represented in the machine and how they are manipulated arithmetically in the central processor.

Binary Representation and Numerical Analysis

Perhaps the best-known quality of the computer is that it operates with binary arithmetic: it represents and manipulates numbers in base 2 (the system that uses only the digits 0 and 1, in contrast to our standard decimal system, which uses the digits 0–9). Of course, our decimal system of representation is pure convention, and any integer we can write in base 10 can just as well be written in base 2, although it may require more digits. For example, 503 in base 10 becomes "111110111" in the more limited system. (Fractions are more complicated; each base finds some easier to represent than others.) Numbers can just as easily be added or multiplied in one base as in another, provided we know the corresponding tables.

Most laymen do not realize that there is no profound reason why digital computers are binary devices. It is not that the number 2 has some mystical significance or even some special mathematical property; computers are binary as a result of our level of technological development. Computers evolved from machines that used hundreds or thousands of telephone relays to perform their calculations, and relays are switches that are either on or off, permitting current to flow or preventing it. Thus, the on or engaged state could represent the digit 1 and the off state the digit 0. When electronic components, first vacuum tubes and then transistors, replaced these mechanical relays, a new form of representation was born. Digits were no longer embodied by open or closed switches but instead by two different levels of voltage; calculations were no longer performed by the snapping of relays but by the much faster changes of voltage in fully electronic circuits. Still, engineers continued to construct two-state devices because, in order to assure reliability among thousands of components, it was best to limit the voltage levels to two. If precise and reliable circuits using five or ten states could be manufactured, then computers could count up to 5 or, as we do, to 10.

The fact that computers today are binary points up another, more fundamental fact—that these machines physically embody numbers and the operations of arithmetic. The status of numbers themselves, of laws of arithmetic, and of theorems of mathematics has never been settled once and for all, and never will be, for it is a philosophical issue about which each culture, perhaps each generation, comes to its own conclusions. Most ages have agreed, however, that there is a radical distinction between number and numeral, between the number 5 and any means we have of representing that number in the world of experience. To work with numbers, mankind has found it necessary to invent symbols for them: pebbles, the marks of a stick in the sand or ink on paper, the beads of an abacus, and now, in the computer age, voltages in circuits and other electromagnetic means. One representation is philosophically as good as any other.

What makes the computer's representation special is that it can be manipulated so rapidly without direct human intervention. Once the program is determined and the machine set to work, the electrons fly until an answer is produced. An abacus can produce an answer mechanically by means of a person who unthinkingly slides the counters according to the rules. And yet the very fact that a human being is needed to push the counters suggests a close link between man and machine. The abacus is a tool rather than a machine, for it extends human technical capabilities while remaining intimately under human control. A machine runs more or less under its own control, with its own sense of purpose and its own inanimate source of power. A human being must provide the logical as well as the physical power needed to make an abacus function, the logical power being the series of rules he applies as he moves the counters. In contrast, the digital computer "moves its counters" electronically, and it keeps its rules stored in its own memory. In the computer, numbers and arithmetic operations seem to assert their own existence in the real world and outside of men's minds, an existence that they never before possessed.

Suppose I want to construct a circuit to add two single-digit binary numbers. The possibilities are limited, and the circuit can be made quite simple (see figure 4-1). The adder may be built out of transistors so that it will alter the voltages in the desired way. For this, there must be two input wires, one corresponding to each addend, and two output wires for the possible two-digit result. The input wires are set at the desired voltage (low voltage for 0, high for 1), and because of the arrangement of the electronic components within the adder itself the output wires assume levels that represent the answer. These new levels can then be used to trigger other operations in the machine. Although it is very fast, this adjustment of voltages, like any physical process, is not instantaneous. As the voltage change occurs in the input wires, the circuit is distorted, and a tiny fraction of a second later the output wires respond with the proper answer. In fact, the computer must wait to ensure that the answer is settled. This jiggling of electrons in time and space is the physical embodiment of addition.

Jiggling electrons are an alternative to the human method of drawing the answer mentally from memorized tables. Building such an adder is not mysterious: it means arranging familiar electronic components according to familiar laws of electrical engineering. The mystery, if there is one, is that electrons should behave as they do, jiggling around in metals and semiconductor materials at tremendous speeds and following laws of attraction and repulsion that make them ideal counting stones for an electronic abacus. Genius was needed to discover these properties of electrons and to give them a mathematical formulation, and considerable skill is needed to build transistors and assemble them into the circuitry of an adder. Yet once the computer is humming, no human has to intervene in the addition itself.

This one small arithmetic circuit can serve as a component in an adder that operates upon larger binary numbers; the output of one adder becomes input for the next. All the mathematical talents of the computer are built from such simple elements. Subtraction can be treated as addition of negative numbers, multiplication as repeated addition (multiplication by powers of 2 as simple shifting of digits), and division as repeated subtraction. In general, the more sophisticated operations are performed as combinations of arithmetic ones. In terms of the circuitry itself, the hardware, the computer seldom knows more mathematics than a twelve-year-old. It cannot directly find square roots, differenti-

Figure 4-1. Binary Addition

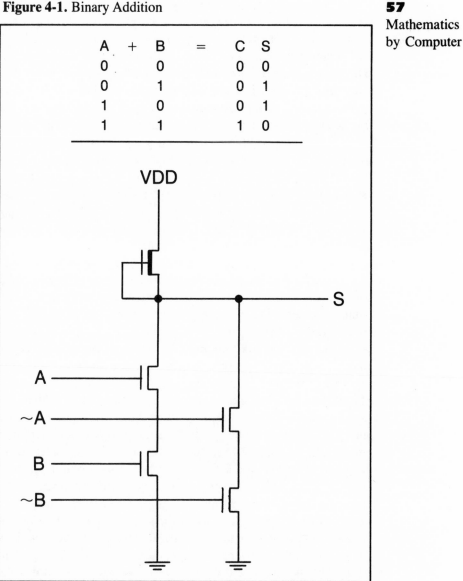

A	+	B	=	C	S
0		0		0	0
0		1		0	1
1		0		0	1
1		1		1	0

The table (above) represents one-digit binary addition. The input values are A and B. The two output digits are C and S. The schematic (below) represents part of one possible implementation of this addition table in actual electronic components. This is embodied mathematics—expressing abstract relationships of numbers as voltage differences in electronic circuits. The S digit output is based on the representation scheme in Carver Mead and Lynn Conway, *Introduction to VLSI Systems* (Reading, Mass.: Addison-Wesley, 1980); © 1980 by Addison-Wesley, reprinted by permission.

ate, integrate, or determine statistical values. These tasks must
be broken down into a sequence of arithmetic operations: a human programmer spells out step by step what needs to be done.

Computer mathematics is not at all as the layman imagines it.
He regards the machine as Charles Babbage hoped it would be,
an infinitely precise calculating engine, grinding out answers
without the possibility of error and so quickly that no problem is
intractable. In fact, computer mathematics is a science of limited
resources—limited time and limited space for calculation—and
by its very nature imprecise and error prone. Computing the solution to a mathematical problem is the antithesis of doing "analysis" in the tradition of the great Western mathematicians from
Newton to Hilbert. The most obvious difference, writes the computer mathematician R. W. Hamming, is "that [traditional] mathematics regularly uses the infinite, both for the representation of
numbers and for processes, whereas computing is necessarily
done on a finite machine in a finite time. The finite representation
of numbers in the machine leads to roundoff errors, whereas the
finite representation of processes leads to truncation errors" (*Introduction to Applied Numerical Analysis*, 2–3).

Numbers in the machine must be represented as strings of binary digits; such a presentation is exact for integers and some
fractions, but many, infinitely many, real numbers (such as the
square root of 2, pi, and the natural logarithm *e*) cannot be represented exactly by any finite string of digits. And any computer
has only a finite number of digits at its disposal, for each digit
must occupy a data cell (made of transistors or some other costly
device). The limitation inherent in every digital machine, its finitude, leads in turn to errors both in representing numbers and in
operating upon them. In most cases, the computer cannot even
add or subtract perfectly; any process more complicated (one of
real interest such as differentiation) can result in errors so serious
as to vitiate the answer. The computer, the least physical of all
machines, is still tainted by its terrestrial origins. When it seeks
to enter the ethereal world of mathematics, it produces only imperfect results. Computer mathematicians spend most of their
time devising methods to keep errors minimal. Their craft is
called *numerical analysis*. Analysis is more or less the art of solving problems using calculus; the adjective "numerical" means
that the solution must be found in concrete terms, in terms of
numbers rather than variables and with techniques that can be de-

scribed by an algorithm and so executed by computer. Numerical
analysts devote the rest of their effort to conquering time, the other limitation of the machine. Despite the computer's remarkable speed, there are many problems in mathematics that the fastest machines would need hundreds of years to solve. All problems must be constituted or restricted, where possible, to come within a time scale that the computer can handle. Even when the numerical analyst has found an algorithm that is not crippled by error, he may not be able to enjoy its full power or accuracy because it is too time-consuming, perhaps even infinite. Accuracy and efficiency are related: the more accurate method is often the slower method.

With all these marks against them, it is fair to ask why computers are used in mathematics at all. The answer is that the machines are set to work on problems (usually involving differential equations) that cannot be solved readily or at all by pure, nineteenth-century techniques. Differential equations are, for better or worse, one of the scientist's chief tools for understanding the natural world in such areas as radioactivity, changes in the weather, biological growth, electricity and magnetism, and so on. The imprecise digital computer, then, is the only way open for the solution of problems that engineers, physicists, or chemists confront daily. The ubiquity of numerical problems for peace or war has made computers indispensable in the world of applied mathematics. And the computer has in turn presented that world with a particularly compelling form of embodied mathematics, a science of discrete and finite numbers and of operations prone to error.

Mathematics and Culture

The next question is why this change in the thinking of mathematicians should have any larger significance. In fact, mathematics has always been an indication of a culture's character and direction. Admittedly, only a fraction of the educated in any age gets very far into the mysteries of mathematics; fewer still can make creative use of the advanced theorems. But there is such a thing as a feeling for numbers, an attitude toward mathematical objects, and this can be achieved without sophisticated training. A feeling for numbers enters broadly into science, technology, philosophy, and even art: it is an important stroke in the cultural signature of any society. Mathematics is not simply a monolithic body of knowledge that has been accumulating since the time of the Greeks or the Egyptians or the Babylonians. It also has a historical dimension, for each age creates its own brand with its own defining problems and interests. The Greek philosophy of numbers differs from the Western European, which differs again from the new electronic philosophy. All have had an impact far beyond the technical sphere of mathematics. A Doric temple is a visual expression of the Greek idea of geometry, a Bach fugue gives voice to the Western love of abstract, algebraic relationships, and the computer age in its turn promises a new and influential numerical philosophy.

It certainly promises a break with the practice of mathematics in Western Europe up to the end of the nineteenth century. That era culminated in the invention of "infinite analysis," the recasting of descriptive geometry (inherited from the Greeks) and algebra (inherited in part from the Arabs) into a new science. Analysis was the study of mathematical functions, which could be interpreted graphically as relations between sets of numbers. It was also a science of infinity, with its exploration of infinite sequences and series of points and numbers; it achieved its first fruits with the invention of differential and integral calculus in the late seventeenth century. Newton himself, who helped to give analysis its modern meaning, had to overcome the age-old prejudice, also inherited from the Greeks, against the concept of infinity. He had to show that infinite sequences were every bit as worthy a mathematical subject as the venerable science of algebra, that "the Reasonings in this are no less certain than in the other; nor the Equations less exact" (from *De analysi*, quoted by Carl Boyer, *A History of Mathematics*, 433). In this area of science as elsewhere, Newton was a leader, and his exact, if infinite, equations set the tone for a new mathematical age. Infinite analysis became the principal study, the defining mathematics of the eighteenth and nineteenth centuries. Strenuous and abstract thought was called for to find the limit of a series or to solve a differential equation. And these mathematical tools, particularly the calculus, could also be applied with spectacular success to problems in physics, as Newton's *Principia* showed. It was the calculus that allowed him to formulate his laws of dynamics, describing with equal clarity the orbits of the planets and the behavior of cannonballs here on earth. It seemed that if, as Galileo had said, the book of nature was written in the language of mathematics, then Newton and his followers had at last begun to understand the grammar of that book.

There was a clear link between this mathematics and physics and the mechanical-dynamic technology of the same period. Analysis was an abstract science that dealt with ideal relations and infinite quantities for which the real world provided only imperfect examples. Yet it was also, mysteriously, a source of power in the real world; Western technology made direct use of this science in its quest to harness and control nature. Newton's physics created an intellectual atmosphere in which technology could prosper, and the two became such constant companions that the phrase "science and technology" eventually became hendiadys. As Lewis Mumford points out: "Machines—and machines alone—completely met the requirements of the new scientific method and point of view: they fulfilled the definition of 'reality' far more perfectly than living organisms. And once the mechanical world-picture was established, machines could thrive and multiply and dominate existence" (*Technics and Civilization*, 51).

Technologists too thought in abstract terms in their pursuit of new sources of power and their mechanization of craft techniques. Their machines evolved further and further away from humanly intelligible dimensions, and they measured this evolution in invented units such as the horsepower and the watt, units that themselves depended upon the abstract concept of time inspired by the mechanical clock. The analytical mathematician exploring the properties of infinite series and the engineer designing the ever more powerful steam engines were manifestations of the same spirit.

Going back further in time, to ancient Greece, one finds a mathematical spirit altogether different from the Western European, one ironically closer to that of the computer age. For, unlike the analysts, Greek mathematicians thought essentially in geometrical terms—terms provided by the synthetic or pure geometry of Euclid's *Elements*, the art of the compass and the straightedge. The abstract relations between numbers are the analyst's prime concern, whereas the geometrical or graphic aspect is only a learning aid. Students of calculus are told at the outset that a graph, a mere picture, proves nothing. For the ancient Greek mathematician, however, this mere picture was far more important; indeed, the proof of a theorem was the ability to draw with straightedge and compass the required geometrical figure, proof "by construction."

The Greeks actually conceived of numbers in terms of geometrical figures. When the ancient followers of Pythagoras, who turned mathematics into a religion, called number the constituent material of the universe, they meant not merely that the universe obeyed mathematical principles but that number itself was material. This view was utterly antithetical to that of Galileo and the moderns. The Pythagoreans identified numbers with geometrical shapes, from which the elements of the world could be constituted. Perhaps under Pythagorean influence, Plato himself came up with such a scheme in his cosmological dialogue, the *Timaeus*. Four regular polygonal solids, all composed of triangles, were the four elements of fire, air, water, and earth. Plato's world was literally made of triangles.

Greek thinkers in general had a peculiarly concrete view of mathematics; their word "geometry," after all, originally meant the art of "measuring the earth," an art that arose presumably from the needs of the surveyor. Yet they combined this view with a love of mathematics as purely rational thought. Plato was proud of the fact that nothing could be less utilitarian than higher mathematics. The value of studying this science was that it drew the student away from the world of the senses, providing a bridge from the concrete to the abstract and perfect. Socrates' interlocutor in the *Republic* has come a long way from the root meaning when he defines geometry as "the knowledge of the eternally existent."

This knowledge must not be identified with the Western analyst's search for infinity, however. Greek mathematicians, who sought to bring neat, definite ratios to bear upon any problem, took a jaundiced view of the infinite. The discovery of the irrationality of the square root of 2 devastated the Pythagoreans because they had found a number that could not be expressed as the finite ratio of two integers. The sect jealously guarded this awful secret, so the story goes, and the unfortunate Hippasus, who let out the news, was drowned by the gods as a punishment. Since the Greeks thought of numbers in geometric terms, the idea of irrationals, which stretch out to infinity, disturbed the clean lines of their geometry. Draw a square with each side of length 1; then the diagonal has the length of root 2, and the geometer is presented with two lines (side and diagonal) that bear no definite relationship to one another. This was more than the Greek

"common-sense" approach to geometry could easily tolerate. As
Boyer remarks of Euclid: "His premises are the dictates of sen-
sory experience. . . . Such purely formal, logical concepts as
those of the infinitesimal and of instantaneous velocity, of infinite
aggregates and the mathematical continuum are not elaborated
either in Euclidean geometry or in Aristotelian physics, for common sense has no immediate need of them" (*History of Calculus*, 47).

A distaste for the infinite was characteristic of much Greek
philosophy as well. Plato and Aristotle regarded the unlimited or
infinite as aesthetically, philosophically, and even morally repug-
nant. To appreciate the Greek attitude, we need only stand before
a Doric temple, whose geometry is so pleasingly complete, self-
contained, and finite. The limit, or outer boundary, of an object
was crucial to the Greek aesthetic sense: what lacked a final form
was incomplete, imperfect. Both Plato and Aristotle believed that
the universe itself was finite; otherwise, it would have been un-
worthy of the name "cosmos," or ordered world.

Here too there is a link between ancient mathematics and ancient technology. As already pointed out, this technology never
abstracted energy by defining it in mathematical terms and never
removed it far from human and animal sources. The ancient
craftsman did not constantly strive to increase the power at his
disposal, as his Western counterpart did. He concentrated instead
on clarity of line, on perfection of form, within the framework
of his craft tradition. The ancient potter at his wheel worked
in much the same spirit as the ancient geometer, who drew
his figures in the sand with two simple tools. In the Athenian
vase or, more grandly, in the Doric temple, the attention to precisely delimited lines corresponded perfectly to the Pythagorean
view of geometrically realized numbers. If the Romans sometimes built on an extravagant scale and without the same precision, we must remember that there was not a single important
Roman mathematician.

Embodied Mathematics

The urge to describe the world in mathematical terms is nothing
new; it dates back at least as far as the Greeks. But for the
Greeks, numbers themselves were concrete, almost palpable
things. It was the Western European mathematicians who first

sought to quantify nature through the abstract sciences of algebra and calculus. Now the mathematics of the computer is changing the terms in which nature is quantified. It is a development out of and away from the work of the last three centuries—a new kind of analysis. It is also a return to mathematical preferences that are much older, to the Greek notion of number as embodied in the real world. For the ancient mathematician, the world itself was composed of geometrical elements; for the computer mathematician, however, numbers are embodied in only a fragment of the world, within the cabinet of a digital computer. But within this tiny cosmos, numbers possess a life of their own. They rest in the core memory waiting to be called upon, they move into the central processor, combine with other numbers, and move back into memory. They impress us constantly with their reality as they spin out answers to our queries.

Embodied mathematics is in both cases finite and discrete mathematics. The numerical analyst is always working against the limitations of his machine, against shortages of time and space and loss of precision. These limitations are the very challenge of his craft: to devise an algorithm that computes with less memory, more speed, and less error. The computer is a much more subtle and intricate device than the compass of the ancients, but the goal is the same—to work toward perfection within precisely defined limits. The computer specialist has as little use for irrational numbers as the Pythagoreans had, although he does not regard such numbers with distrust or religious awe. On the other hand, if infinite ratios and infinite precision could be introduced, his work would be easier, so much easier that it would perhaps lose some of its appeal and certainly all of its present character.

Computational mathematics is not, of course, a return to the descriptive geometry of the ancients because analysis is still fundamental to the scientific project that began in the seventeenth century and continues today. Mathematicians and engineers are not going to forget how to work calculus problems by analytical methods; indeed, they use those methods every day. Meanwhile, pure mathematics has diversified into many areas, such as topology and number theory, that resist computerization. Still, the major theoretical work in real number analysis was completed by about 1900, and it is now the engineer's concern to turn theory into practice that fills the pages of journals of numerical analysis. What these numerical analysts have in common with mathematicians of the nineteenth century is the theory behind their work.

What they share with ancient geometers is a feeling about the nature of numbers and the efficacy of mathematics in the finite world of our experience.

There remains some distaste among contemporary pure mathematicians for this "dirty," finite, inelegant mathematics by computer. Recently, the famous four-color-map theorem was proved with the aid of a computer. The theorem states that any map in a plane may be colored with at most four colors so that no two countries with a common border have the same color. The problem was one that any layman could understand, but it had defied solution since the nineteenth century. Many mathematicians had contributed to narrowing the problem, but none could frame a final elegant proof. The two authors, Kenneth Appel and Wolfgang Haken, used traditional proof techniques to show that some 1500 maps represented all the possible types that could be drawn; they then called upon the computer to test these maps. There had been some question whether even a fast computer of the mid-1970s was up to the task, and by the end they had used 1200 hours of time on three different machines. The authors have suggested that there may be no elegant formal proof of the four-color theorem but only the trial-and-error method the computer makes possible, the exhaustion of a finite number of cases. They conclude: "We believe that there are theorems of great mathematical interest that can only be proved by computer methods" ("Solution of the Four-Color-Map Problem," 120).

Their success points to a further rapprochement between pure mathematics and the computer, but the present collaboration of man and machine is already significant. Numerical analysts are technicians who have learned to work within and indeed through the limits of the new embodied mathematics. This working through limits is characteristic of all who use the computer and of the machine itself.

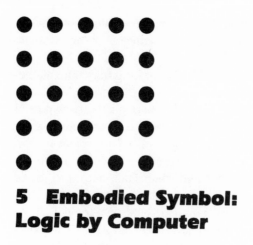

5 Embodied Symbol: Logic by Computer

The computer was built to solve mathematical problems, but it was soon realized that its power went beyond numerical calculation. It could manipulate arbitrary symbols as easily as it could add numbers. Lady Lovelace, the disciple of Charles Babbage, understood the power of machine computation in the last century when she remarked that the Analytical Engine "can arrange and combine its numerical quantities exactly as if they were *letters* or other *general* symbols; and in fact it might bring out its results in algebraical *notation*, were provisions made accordingly" (Morrison and Morrison, *Charles Babbage and His Calculating Engines*, 273). John von Neumann had come to the same understanding when he arranged for his machine to keep the programs in memory along with the numerical data and to keep them in the same binary format. The program was a coded list of commands, a binary representation of the sequence of actions to be performed by the computer in reaching an answer. The computer did not calculate these commands, rather, it interpreted and executed them; it executed commands and added numbers with equal ease. The von Neumann computer was not a calculator but a logic machine.

The machine demonstrated the unity of all systems of representation: as symbols, numerals are no different from letters, hieroglyphs, or ideographs, for all these can be manipulated in electronic circuits. It soon became clear that the manipulations of mathematics were not fundamental to the machine. Computer mathematics itself could be defined in terms of the science of symbolic logic, a science that had been developed by generations of philosophers in the nineteenth and twentieth centuries. But symbolic logic too underwent a change when it was adapted for the computer: like mathematics, it took on the peculiar qualities and limitations of the machine.

Truth and the von Neumann Machine

There is a great philosophical issue that the computer requires its programmers to confront from the first day they begin to learn their craft—the nature of symbols and symbolic representation. To solve a problem, the computer must fashion some representation of the problem within its circuits and then operate, and the programmer must always be aware of the relation between the original problem and the computer's means of representation. He must constantly attend to the symbols that stand for facets of the problem, and he is constantly impressed by the way these symbols are threaded through the processor and manipulated as they are transformed into the desired values of the output.

The computer teaches forcefully the lesson that symbols are arbitrary, that they mean exactly and only what the programmer and the machine define them to mean. This is a lesson that might also be learned from the world of natural language, as a colleague has suggested to me. The meaning of the word "inhabitable" seems clear: "habitable, capable of supporting life." This combination of letters is not particularly ambiguous and is not a homonym. But present the same letters to a Frenchman, and he will interpret the word in almost the opposite sense. The alphabet is the same for French and English, but the words made from that alphabet are different. There is no intrinsic meaning to the string of letters "inhabitable"; the meaning we assign is as valid as the one assigned by the French. But in order to know the meaning, we must know which "code" (language) we are using.

The computer operates with a variety of codes, but the genius of the von Neumann scheme is that the codes can all be expressed in the same binary alphabet. The computer represents numbers, letters, and programming instructions all as strings of *bits* (binary units of information). Just as there is a binary representation for the number 12, so there is one for the letter "a" and one for the command to copy a value from the CPU into the memory. The instruction code (the *opcode*) is a set of symbols that command the processor to perform its operations: to add two numbers, to store a value in memory, and so on. Then there are codes that the machine uses internally (in its CPU and memory) to represent numbers for calculation. All coded values ultimately exist only as arbitrary configurations of bits. Inside the computer, the same bit string might represent the number 115, the letter "s," and the opcode that instructs the CPU to add two numbers. How the string is treated—as number, letter, or instruction—depends generally on the context: that is, the machine itself interprets each string by its relation to other strings in the memory.

Now it may seem odd that letters and punctuation should need representation in the electronic world at all. The string "01110011", interpreted as the number 115, can sensibly be added to another string standing for another number, and so the processor can proceed through a series of meaningful calculations; but it makes no sense to add two strings that represent letters. An operation that does make sense is to compare two letters or sets of letters forming names. By simple comparisons, the computer can be programmed to search through a list of names or to sort a list into alphabetical order. The prosaic operations of sorting and searching are basic to the nonscientific uses of the computer, especially business uses, and the success of such applications has provided the resources and much of the impetus for more esoteric programming. The most sophisticated programs for playing chess still rely heavily upon the two techniques used by the telephone company in billing its customers.

There is more that can be done with strings of nonnumeric data; in fact, these strings can be "added" in a meaningful way. A two-state system (on/off, high voltage/low voltage) may be used to denote yes or no, true or false. This logical interpretation opens for the computer a new universe of discourse, as logicians are fond of calling it: any statement that logicians regard as true or false can be represented as a bit in an electronic circuit. Such bits become the numerals of logical arithmetic, and the operations of this arithmetic are called "truth functions." A complex science has been built from this notion.

Take the statement: "Von Neumann died in 1957." This statement is true and earns the truth value 1. Its negation, "Von Neumann did not die in 1957," is false and has the value 0. Logicians would assign a symbol, A, to the original statement and say that

A has the value 1; the negation of A, written ~A, has the value 0.
Now consider a statement that is false: "Babbage was born in
1903." If this statement is called A, then the formulas are reversed: A has value 0 and ~A has value 1. This exhausts the possibilities for the simplest truth function, the *not* function, which
inverts truth values according to the following table:

A	~A
0	1
1	0

The table expresses the trivial notion that if a statement is true,
then its negation is false, and vice versa. The point is that by reducing this notion to a formula and representing truth as a binary
digit, we make the realm of symbolic logic accessible to the computer. Here the machine is not adding numbers; it is determining
truth and falsity, albeit in the precise and limited sense of the logician. Other truth functions can be defined by more complicated
tables—an *or* function, an *and* function—and soon the whole
complex structure of formal logic can be expressed in the computer's alphabet.

This highly mathematical treatment of truth and falsehood,
which the computer programmer now uses daily, was created by
philosophers and mathematicians in the nineteenth and early
twentieth centuries, before computers had any more concrete realization than the unassembled pieces of Babbage's Analytical
Engine. These thinkers had in mind another goal altogether: to
free European logic from the hold that Aristotle had exerted since
ancient times. Their complaint was that Aristotle had divorced
logic from mathematics. By making logic rigorous and mathematical, they now hoped to put it at the center of European philosophy, the position it had always promised to occupy and yet
never quite achieved. They felt that a logical calculus (on the
model of the rigorous mathematics of nineteenth-century analysis) ought to be the philosopher's main tool in his search for truth.
So they created the *formal systems* of propositional and predicate
calculus, in which symbols without any intrinsic meaning were
manipulated according to universally valid rules of thought, rules
about contradiction, consistency, and implication.

The *Principia Mathematica* of Bertrand Russell and A. N.
Whitehead (1910) was nothing less than an attempt to make symbolic logic the foundation of mathematics and so to provide a
bridge between philosophy and the most rigorous of sciences.
Russell himself elsewhere wrote that in his day "logic has be-

come more mathematical and mathematics has become more logical. The consequence is that it has now become wholly impossible to draw a line between the two; in fact the two are one. They
differ as boy and man: the logic is the youth of mathematics and
mathematics is the manhood of logic" (*Introduction to Mathematical Philosophy*, 194).

The project to unify logic and mathematics was the height of
abstraction, as a glance through the pages of the *Principia* shows,
and by its very complexity and rigor, the new symbolic logic
placed itself beyond the reach of many philosophers and most
other intellectuals. The Aristotelian categories and classification
of syllogisms could be understood even by humanists. Everyone
knows the example:

> All men are mortal.
> Socrates is a man.
> Therefore, Socrates is mortal.

In symbolic form, the syllogism is comparatively easy to follow:

$$(x) (Hx \rightarrow Mx) \qquad \text{where } Hx = x \text{ is a man,}$$
$$Hs \qquad Mx = x \text{ is mortal,}$$
$$\therefore Ms \qquad \text{and } s = \text{Socrates.}$$

But once we have made the leap to symbolic representation, we
begin to reason entirely abstractly (without reference to men,
mortality, or Socrates) and in a layered fashion, building one
level of complexity upon another. The syllogism above stands in
relation to the formal reasoning of the *Principia* as arithmetic
stands in relation to theorems of complex variable analysis—it is
a bare beginning. Few of us indeed can follow an argument of
mature symbolic logic. For most, the incentive to learn this science was never adequate, and in any case the edifice that Russell
and others had erected failed decisively in the 1930s. The work of
Kurt Gödel showed that there were inherent limitations to any
logical system designed for constructing mathematics: the danger
of contradictions or lack of completeness in such systems could
not be eliminated. Symbolic logic remained a compelling but esoteric discipline of its own, influential but still too abstruse to direct the course of twentieth-century thought.

The invention of electronic computing hardware provided a
new and unexpectedly utilitarian sphere for symbolic logic. It
was not a revivification, for there were important logicians
throughout the first half of the twentieth century. Suddenly, how-

ever, their work came down to earth when logical truth functions
came to be embodied in electronic circuits. Logic and electronics
married, and the offspring was the von Neumann computer. For
example, put a logical truth table beside its circuit diagram (figure 5-1). The difference between the circuitry and the truth table
may seem to the layman to be merely the replacement of one set
of symbols with another, but there is more to it. In one case, the
symbols are those of logic, counters that have meaning only as
we define them. In the other, the symbols stand for electronic
components, which are built from such earthly materials as silicon and gold (the most pedestrian and the most princely) and set
to work in the world of experience.

The central processor of the computer performs its calculations
by means of just such circuitry (*and* gates, *or* gates, and the like),
and engineers speak of designing the "logic" for a particular machine, the constellation of logical circuits the machine will use.
Such circuits exist throughout the computer but are concentrated
in the ALU, that part of the central processor through which units
of data are sent to be added, compared, and otherwise manipulated. There need not be two separate sections of the ALU, one
for arithmetic and one for logic, because the arithmetic operations can be performed by the same circuits that embody truth
functions. A binary adder, for example, can actually be built using *and*, *or*, and *not* gates, which is simply another facet of the
remarkable unity of design of the digital computer. Just as all
numbers and other symbols are reduced to the same binary digits,
so all arithmetic operations are reduced to the logical manipulation of these digits. And since computer mathematics is nothing
in the end but combinations of arithmetic operations, all the
mathematics of the computer age (including the most complex
statistics and differential equations) depend upon the realization
in circuits of a few elementary logical functions.

The Triumph of Logic

In the computer, then, symbolic logic has achieved what it could
not achieve in the cryptic pages of Russell's *Principia*; it has become the foundation of computerized mathematics. But to win
this end, logic has had to condescend to become concrete, to
clothe itself in a mantle of free-flowing electrons and to preside
over an equally concrete form of mathematics. This is an as-

Figure 5-1. Truth Table and Circuit Diagram

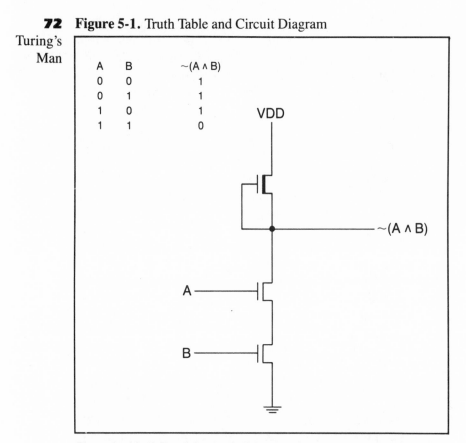

A	B	~(A ∧ B)
0	0	1
0	1	1
1	0	1
1	1	0

VDD

~(A ∧ B)

A

B

The truth table (*left*) and the circuit diagram (*right*) are two representations of the statement: "A and B are not both true." But there is a profound difference. The truth table is simply a logician's code; the circuit diagram stands for the physical embodiment of this logical idea. A digital computer is the embodiment of the whole realm of symbolic logic. The circuit is a NAND gate drawn for nMOS technology, as it appears in Carver Mead and Lynn Conway, *Introduction to VLSI Systems* (Reading, Mass.: Addison-Wesley, 1980), 15, fig. 1.18; © 1980 by Addison-Wesley, reprinted with permission.

tonishing development in the modern history of ideas. Its impact has not been fully appreciated, perhaps because symbolic logic in its pre-electronic form was not available to the intellectual community at large. Now, however, the most esoteric branch of philosophy is being put to a practical, often mundane use, and the most practical of people, engineers, are studying the rudiments of symbolic logic. This pragmatic feeling for electronic logic has worked its way into the bones of Turing's men; it is a key part of their world view.

Logicians have often dreamt of reducing (or raising) all rational human thinking to the level of their craft. Aristotle, it is true, did not think that all fields could be reached by the syllogistic logic he had invented. It would be as unreasonable, he declared in the *Nicomachean Ethics*, to expect exact proofs from a politician as to accept merely probable conclusions from the mathematician. But later logicians, again with characteristic Western enthusiasm, have indeed hoped to attain mathematical perfection in religion or politics. Leibniz himself proposed a universal language in which all simple ideas could be represented by individual symbols and then combined by the rules of logic; when expressed in this language, even the truths of religion would be irrefutable.

With the computer, this hope is resurrected in a new guise. Any task the computer performs is a matter of the logical manipulation of symbols; each new problem solved is a conquest for the logical calculus of thought and gives encouragement to those who still follow Leibniz's program. "If I were to choose a patron saint for cybernetics [the study of living organisms as logical machines] out of the history of science," wrote Norbert Wiener in 1948, "I should have to choose Leibniz. . . . The *calculus ratiocinator* [calculus of reasoning] of Leibniz contains the germs of the *machina ratiocinatrix*, the reasoning machine" (*Cybernetics*, 20).

When a computer specialist speaks of his machine "thinking," "reasoning," "manifesting intelligence," or "solving problems," he means that it is operating according to the rules of its embodied symbolic logic. There seems to be no bias attached to electronic logic, as there is still to some extent to electronic mathematics. It is instead a triumph that theory can now be put so successfully into practice. After all, if formal logic is now used to sort bills and keep inventories, it is also invaluable in exploring the planets with unmanned spacecraft, deciphering data collected in particle accelerators, and, least important perhaps but most dramatic, in playing computerized chess. Many computer specialists believe that no important intellectual endeavor will stay forever outside the scope of their calculus of reasoning.

The computer is also a triumph of "logic" in another sense, the popular sense, suggesting the compulsion for tight constructions, careful definitions, and tidy arguments. The computer is the embodiment of the world as the logician would like it to be. The

world of everyday experience has the annoying tendency of spill-
ing out of the neat categories he constructs, and it has had this
tendency ever since Aristotle codified logic in the fourth century
B.C. For the logician, then, the digital computer has this virtue:
its design is perfectly logical down to the scale of electrons; it has
conquered the disorder of the natural world by the hierarchical
principles of symbolic logic. Outside the computer electrons fly
about obeying no final cause but only the statistical and quantized
laws of physics. According to the principle of entropy, the universe as a whole moves toward greater disorder. Meanwhile, the
thoughts and beliefs of men—to the philosophically minded logician, entities as real as tables and chairs—are equally chaotic.
No philosopher has yet produced a rigorous proof for or against
the existence of God that could satisfy more than a fraction of his
colleagues. But in the world of the computer, a special kind of
order exists. Electrons still obey the statistical laws of physics,
but the circuits are so designed that these laws themselves serve
the principles of logic.

Western technology has been based on the idea of subordinat-
ing nature to artifice, and it has been a more or less qualified
failure. Natural laws always prove stronger than human constructs; machines are inaccurate and eventually, usually all too
soon, they break down. But engineers have never had at their disposal a tool of design as precise as symbolic logic. They have
never attempted to master nature on so fine a scale. Electronic
technology is mankind's most arrogant and most successful attempt to impose its own teleology upon the natural world. The *or*
gate in a computer is really the final cause for which electrons fly
about, a final cause dictated by the engineer who designed the
machine. The logical world the engineer creates reminds us in
this respect of the universe of a Greek cosmologist. The logic of
Aristotle has been left far behind, but the typically Greek contrast
between order within and chaos without remains. For Greek
thinkers, the universe was a place of order; cosmos, in fact,
means "order." Beyond it there may exist nothing at all or total
disorder, which is the same as nothing. The boundary between
the two is all important: it is precisely where Aristotle placed his
god. In a similar way, the lines of a computer circuit diagram
draw a map of an ordered world. Outside the computer, there is
only chaos, dust, and various contaminants from which this frag-
ile universe of order and logic must be guarded if it is to continue
to function.

Electronic thought is the process by which the computer fires small amounts of data back and forth through the circuits in the central processing unit. Just as the popular imagination envisions, the computer "thinks" by means of dispassionate, logical calculation. A thought in common language can be anything from an image in the mind to an emotionally charged point of view, an abstraction like justice, or a theoretical entity invented by physicists. Yet dreamy imaginings, feelings, ambiguities, and contradictions have no meaning in the central processor of a von Neumann machine. The anomaly of half-completed thoughts, which come sometimes as we are falling asleep, also cannot be computed. In binary logic, a bit is either off or on, false or true.

There are four qualities that characterize the logical processing carried on by the CPU: it is discrete, conventional, finite, and isolated. These qualities properly belong to all formal systems, and before about 1950 such systems existed only in the pages of the logician's notebook. Since then, they have been dancing through the circuitry of digital computers. Daily success in using formal logic is suggesting to Turing's man that this is the only way, or at least the preferred way, by which thought (human or electronic) can proceed. These four qualities will emerge repeatedly in the chapters that follow, for they define the new views of time, space, and language that will be explored.

All information, all knowledge, must be coded in some binary representation in order to be acted upon by the computer. It must be broken into a series of discrete values. It has already been noted that even in mathematics this is a limitation, that infinitely many numbers (irrational, algebraic, transcendental) cannot be represented exactly in a digital computer because they cannot be reduced to a finite series of decimal or binary places. Approximation leads to error, the central problem of computer mathematics. On the other hand, symbolic logic has no difficulty adapting itself to discrete representation, for it is characteristic of logic to seek to reduce the continuous to the discrete, the ambiguities and uncertainties of everyday thinking to the binary scheme of true and false. What is remarkable is that the idea of a two-valued truth function should be so well suited to representation in electronic circuits. In any case, there is no room in the computer for the continuous.

The computer is of course often called upon to process continuous data from the physical world. When a spacecraft flies by the planet Mars, the pictures it takes are like television images, made up of perhaps several hundred rows, each containing dozens of individual points. At each point, the camera records a particular intensity of light, and the intensity is converted into a numerical value, say, 1 for very dark and 20 for very bright. The whole picture then is represented by a matrix of numbers from 1 to 20, and it is these numbers that are radioed back to earth. A computer takes the received values and reconstitutes the picture, again made of individual points. The original shadings of light given off by the planet have been lost, but if the resolution is fine enough, the human observer on earth will hardly notice the difference. Newspapers print pictures by a similar technique, using dots to give the illusion of continuous shades and contours. For the computer, the continuous is always simulated by a grid of discrete values. Even recorded sounds can be digitalized, turned into a series of numerical values, and later turned back into continuous wave forms—all with extraordinary fidelity.

Computer thought is wholly a matter of convention, of formal rules acting upon contentless symbols. Whether numbers or letters are represented as bit strings in the machine, the representation is one of pure denotation. The number 10 spelled out in transistors has no connotations whatever, such as the word "two" written or spoken is likely to have. For the ancient Pythagoreans, the word "two" possessed an extraordinary range of connotations, including femininity, darkness, infinity, and evil, which were all somehow associated with even numbers; for moderns, perhaps, it has connotations of harmony and completion. Bits within a computer are logical symbols that mean nothing more than they are deemed to mean in the context of a particular program. (As already noted, the same string can stand for a number, a letter, or a machine instruction depending upon the context.) The programmer, then, creates meaning within his program by convention.

Computer specialists like to speak of their machines as manipulators of symbols. This is a good characterization so long as we remember that the symbols are not literary devices or natural signs pointing to a higher reality but simply arbitrary units that the programmer may choose to interpret as words in the English language or moves in the game of chess. Indeed, chess programs are a perfect example of this quality of electronic thought. A computer playing chess is manipulating a series of conventions

(the rules that determine how the pieces are to be moved), and
computers play good chess only because these conventions can
be clearly stated in the machine's electronic vocabulary and manipulated by the central processor.

There is something straightforward and apparently superficial
about this electronic thought process. It seems superficial because of the long Western tradition of realism, of the conviction
that the human mind does not construct its ideas purely at will but
that instead those ideas have some force, necessity, or reality of
their own. The mind therefore discovers ideas rather than inventing them and may hope through this process of discovery to attain
some higher, perhaps ultimate, knowledge. The computer suggests no such deep correspondence between thought and the
world at large, such as philosophers and poets from Plato to Descartes, from Greek tragedians to French symbolists, have found.
Computer thought is a triumph of nominalism. The symbols that
the computer manipulates are senseless in isolation, for there is
no reality "behind" the symbols, propping them up. Only when
incorporated into a program and set to work, does each symbol
acquire a single, unambiguous function. And the whole program
is viewed as an invention rather than a discovery.

Computer thought is a sequence of operations, of fetch-and-
execute cycles of the central processing unit. Computer thought
is therefore discrete and finite in a second important sense. Not
only are numbers and words represented in discrete strings, but
these strings are also processed discretely—from the beginning
of the program to its termination. Programming shares this quality with the mathematical thought of both classical Greece and
Western Europe, in which the mathematician moves step by step
toward his goal. The ancient geometer made a series of verbal
assertions about his figures, each linked to a preceding assertion
and justified by the rules of inference. His European counterpart,
especially in the nineteenth century, often expressed his assertions in equations or sentences of abstract symbols. But the principle remained that there could be no dramatic leaps in the proof,
that everything should follow as the night the day. Truly creative
mathematicians have always made such leaps in their preliminary
thinking, in their intuitive solutions to problems. In fact, their assertions, inspired guesses really, may go unproved for decades or
centuries. But when the proof comes, it must be justified at every
step. Every intuition must be transformed into a deduction.

Every computer program is the electronic realization, the tangible proof, of a theorem in logic. We saw in the last chapter that the mathematicians Appel and Haken actually incorporated computer programs into their proof of the four-color theorem. Computer specialists such as Edsger Dijkstra have explicitly treated programs as mathematical entities in their elegant work on logical "correctness." Every programmer, no matter how pragmatic or mathematically naive, is a logician with a theorem to prove. The theorem states: given a certain pattern of bits (program and data) as input, the machine will produce another pattern, the desired output. The programmer asserts such a theorem as he loads his instructions into the machine; he asserts that the program works. He claims that his square-root program, if given the bits standing for the number 5, will indeed produce the square root of 5. The commands of the program themselves are steps of his proof, and the test comes in allowing the CPU to thresh its way line by line through the code. The logical organization of the machine (the way it executes commands) is a finite set of rules that the programmer must obey in constructing his proof, just as the mathematician must obey laws of inference in constructing his.

The word "finite" is very important. The pursuit of infinity has been a prime issue for mathematics, philosophy, and even poetry since the Middle Ages. Although the Greeks were repelled by the infinite, Western Europeans in a sense worshiped it. Was not God himself infinitely powerful and good? Should not men and women strive to come as close to God's infinity as their own finite natures would allow? The infinite and the infinitesimal terrified and yet fascinated philosophers like Pascal, and the paradox of the infinite was still identified by Kant as a major philosophical dilemma. If the computer specialist can accept only finite numbers and finite logic, this represents a turning point in the history of ideas.

Finally, computer thought is thought in isolation. It is indeed embodied, but the embodiment remains curiously separated from the rest of the physical world. I spoke earlier of the contrast between the special organization of electrons within the machine and the disorder of the inanimate world outside. The walls that separate the two worlds, those cabinets of aluminum and colored plastic, make sure that, apart from the electric current for power, there is no contact between the computer and its environment. There is a correspondence between the two, for the equations generated by the machine may successfully describe the course of a spacecraft reentering the earth's atmosphere. But this is the tantalizing correspondence between mathematics and the world in general: it is a formal relation rather than an immediate contact. The CPU does have channels by which it can communicate with the outside, such input/output devices as tape and disk drives and television screens. But in order to communicate, these devices must themselves obey rules of representation that the CPU dictates; nothing but electronic signals standing for binary strings may come in or go out.

On the other hand, electronic thought is not introspective in any earlier philosophical sense. The processes of the CPU are open to public inspection; we need only take the cover off the machine and apply electronic probing devices (though this serves no purpose unless the machine is malfunctioning). The computer is isolated in the way that all mathematical and logical thought is isolated: it allows for no possibility of immediate union between two thinkers. Computers can and do work together in so-called networks, and several CPUs can be combined within a single frame. But in the current technology, if one processor were to interfere with the cycles of another, the result would be chaos.

What about the historically important idea that immediate union between minds was possible and perhaps vital to the human condition? Plato believed that the task of the philosopher was to reach the world of ideas by dialectic and contemplation but certainly not by his senses (his input/output devices, to use the computer metaphor), which are mere distractions. His mind must somehow unite itself with the idea of the good and the beautiful. In the same way, many Christian philosophers aimed at some sort of mental union with God. Even the mathematician Descartes had radically separated mind and body and envisioned a community of minds with God. This kind of mental activity—whether the comparatively calm unification envisioned by Plato or the more violent upward movement imagined by mystics and poets of Western Europe—has no counterpart in computerized thought.

The computer does not strive; it proceeds to a predetermined goal. The striving after infinity or self-knowledge or God, so important in the previous age, is especially foreign to electronic thought. And if the Western mind has often dreamed of overcoming its mortal limitations, its finiteness, in one way or another, the programmer has no such ambitions. The very structure of the CPU and of the languages used to program it reminds him that he must proceed a step at a time and in a finite number of steps produce a useful result.

6 Electronic Space

There are three kinds of space of interest to us here: space as human beings perceive it, as they use it technologically, and as they theorize about it. The perception of space is studied by psychologists. The use of space is the subject of geography, sociology, and planning of all sorts—that is, the way in which human beings use tools and techniques to create an area for living and to allocate available resources. Meanwhile, scholars of art and architecture study the artist's use of space for its symbolic or ornamental significance. Finally, there is the space about which philosophers have been arguing since Greek times, debating its nature and indeed its very existence: is space full or empty, and how does it interact with matter? Joining in the debate have been the mathematicians, whose geometry and, more recently, set theory define "spaces" that may not resemble the space of either philosophy or human experience.

At least two of these kinds of space are constantly changing with the changing tools and preferences of each age. It may be that all human beings have the same perception of space at the biological level of perception. But certainly every society uses its space differently, both technologically and artistically. A Greek polis did not express the same habits of town planning as a Renaissance city or a modern suburb, and Greek artists did not use architectural or sculptural space as did Western Europeans. As for philosophical or mathematical space, Aristotle certainly understood it differently than Descartes, Newton, or a twentieth-century physicist would.

Within each culture, craftsmen, architects, sculptors, philosophers, and mathematicians all share a general attitude towards space: what the sculptor expresses in stone, the philosopher may explain in a treatise. But craftsmen, town planners, and even artists were not accustomed until recently to analyzing their manipulations of space. They talked or wrote simply, sometimes poetically, but seldom abstractly about their craft; they relied on intuition more than logic. This was certainly true of the ancients, where the humbler craftsmen were silent. Even the architects of the Doric temple probably lacked the critical vocabulary to explain its remarkable geometry. Nor could a medieval architect have said much about the soaring spaces of a Gothic cathedral. The development of such a critical vocabulary for art is recent, as is the study of the history of technology itself.

Roles that have been expressed separately in previous ages are now unified in the age of the computer. If the computer programmer is primarily a craftsman in his manipulation of computer space, he is also a mathematician or philosopher because of the nature of the materials with which he works. Here is another instance of the duality of Turing's man, of the computer's requirement that he think both in physical terms and in abstractions, pragmatically and philosophically. We shall see that there are two facets to computer space (both of which are called "space" by designers and programmers): *physical space*, properties of the computer as a machine made of terrestrial parts and subject to terrestrial limitations, and *logical space*, properties of the computer as a logical entity, removed as far as possible from its mechanical origins.

Physical Space

Elements of data cannot be manipulated endlessly; they cannot remain forever in the central processor. At some time, they must be deposited, and new data fetched. In addition to the CPU, the second major unit of the von Neumann computer is its storage or memory, the bank of electronic devices in which programs and data reside while waiting to be processed and to which intermediate results are returned. This storage is a collection of cells, each of which can hold one unit of data, and these cells form the physical space of the computer. They are physical devices, occupying space inside the cabinet and using electricity to store and maintain information. Today the cells are most often made of transistors, painted in large numbers onto chips of silicon. In fact, memory is currently the most common and the most lucrative kind of integrated circuit chip. In the past, memory cells were composed of doughnut-shaped cores of ferrite and so acquired the name *core* storage, a term still in use after the technology has been superseded. However they are constructed, these cells can give out or receive information from the central processor very rapidly, in millionths of a second or less.

Like most other components, memory chips are arranged in the computer in regular, even geometric patterns. The chips are placed in rows on circuit boards, the boards are inserted in slots in the machine. Transistors within each chip and the chips on each board are wired together to allow for the organization of stored data, an organization more or less standard in contemporary machines. The bit is the machine's fundamental unit, but it carries little information in isolation. As larger units are needed, bits are grouped into *bytes* (eight bits), and bytes into *words* (often two or four bytes, depending upon the machine). The processor generally fetches and manipulates data in bytes or words. The hierarchy continues with the grouping of words into pages and often pages into segments. For any given computer, these logical divisions could all be traced through the geometry of the memory system down to the level of the very transistors that store each bit.

The physical space of the computer has two familiar qualities: it is discrete and limited. Memory systems are discrete in that each cell can represent only one bit of data; computer space is ultimately composed of discrete data points. Each cell is itself rather complex, made of a couple of transistors with related wiring: it occupies space, consumes electricity, and generates heat. This puts a limit on the capacity of memory, a practical limitation that has applied since the beginning of the computer age and will continue to apply as far into the future as we can now look. Admittedly, each improvement in technology, each new generation of machines, has eased the shortage of available space. The first machines in the late 1940s had only a few hundred or thousand bytes of storage, but today even microcomputers (desk-top models) routinely make available hundreds of thousands of bytes. Still, transistor memory is relatively expensive, and it is never possible to satisfy the programmer's appetite for more room in which to work his solutions. One of the goals of large-scale integration (the placing of more and more circuits on a single chip) is to provide more capacity for the money, and yet each time the engineer provides more memory, programmers decide they can now confront a new set of problems that previously required too much data for even the larger machines. There is no fear of running out of such problems in the near future. Storing all the various moves of the game of chess would require, for a memory somehow employing individual atoms as memory cells, perhaps all the matter in the universe. Lack of space is one of the two principal limitations of the electronic world. The other is computer time. Making intelligent use of the space at hand is a cardinal virtue in the craft of computer programming.

Logical Space

In most machines, every byte or word has a number associated with it, which is called its *address*. The numbers are generally assigned serially, from zero to the capacity of the machine, making each byte selectively available to the central processor. Fast, so-called *random access* is a defining quality of computer memory. The access is not random in the sense of being haphazard, however. The processor knows exactly which portion of data it wants, but to obtain that portion it may go directly to the byte or word in question. If it wants byte 5000, it need not obtain all the data from 1 to 4999 and count its way through. The CPU commands the memory address circuits to fetch the needed information, and that byte is immediately presented to it. Computer space, then, is algebraic by its very design because it consists of a series of numbered points or units.

The numbering or addressing of space derives from analytic geometry and is as old as Descartes; it is crucial for putting the physical memory space at the disposal of the programmer. Computer space becomes malleable, manipulable, because it is addressed. Addressing allows the programmer to get at each cell and to shuffle the contents of various cells as he likes; it allows him to begin to view computer space abstractly. He speaks therefore of the *address space* of the machine—the range of numbers that can be invoked in fetching and storing data. Most programmers know very little about the arrangement and operation of the physical memory of their machine: they do not know how the CPU decodes memory addresses and finds data, nor even what locations are named by what addresses. Usually only the designer of the machine could tell easily on what board or chip byte 20,000 is to be found. Layers of technology (both hardware and software) protect the user from the boring and complicated tasks of memory management. Instead, his attention focuses upon the logical idea of memory, the logical space in which his program and data will reside.

This logical space exists in the programmer's mind, or wherever such mathematical ideas as Euclid's perfect triangles exist. The programmer ignores those qualities that have no bearing on his problem (such as the pattern of wires and transistors forming the memory) and is left with the fundamental qualities of electronic space. Logical space is not computer memory freed of its physical limitations, for the corporeality of information in the computer imposes restrictions and offers possibilities that the programmer cannot ignore—rather, it is computer memory reduced to its essence.

For example, the programmer seldom works directly with numerical addresses, which are notoriously hard to remember. Programming languages allow meaningful names to replace numbers in memory. Suppose the program refers several times to the changing velocity of a spacecraft. The velocity is stored, say, at the location whose address is 8000, and the programmer decides to call this location SPEED. Whenever he wants to fetch the velocity from memory and update it, he refers simply to SPEED in his program. A language such as PASCAL allows this convention and quietly substitutes the number 8000 for each occurrence of SPEED when it translates the program into machine instructions. In fact, PASCAL works so unobtrusively that the programmer is never informed of the true numerical address of his data, and he does not care to know. His logical memory space consists of the names of variables such as SPEED, DRAG, and POSITION, together with mathematical relations among the variables appropriate to these names. Meanwhile, the physical memory holds a series of addressed locations, binary digits representing the values of SPEED, DRAG, and POSITION.

The language through which the programmer commands his machine is selectively opaque; it masks absolute addresses but allows him to see and exploit variable names for representing his problem. In a modern computer center, furthermore, each programmer is presented with a *workspace* or region of storage (measured in bytes) and is allowed to fill it in any logical way

he chooses. Since several programmers may have workspaces at
the same time, the computer system must be careful to keep them separate. In general, boundaries around each program are guarded automatically by the system. One user may be allotted bytes 100,001 to 200,000, another bytes 200,001 to 300,000. If the first tries (usually through error) to fetch byte 250,000, his program will be terminated (aborted) by the system, for he is trying to tamper with another user's space. Inside the sandbox of his own logical region, he is free to read and write data, to manipulate structures as he pleases. Again, he usually knows nothing of the physical locations that embody his or anyone else's workspace. He refers to variable names, and it is the task of the program (or several layers of programming) to find and modify the proper physical locations.

A workspace is a logical entity, yet it can never quite break free of its physical and spatial origins as rows of transistors. Thus, the programmer is inclined to imagine his workspace as an area with physical dimensions. He organizes his space by filling it with data, by giving his data a structure, very often visualized in two or more dimensions, to express the relationships that hold among portions of the data. His program then operates upon this structure and returns a new structure as its result: it manipulates the workspace, changing its shape from one that represents the initial conditions of the data to one that represents the desired output. In this sense, programming is the art of expressing logical problems in geometric or architectural terms, of "building structures" in one's logical space—as if one were an architect—that reflect the problem and lead to a solution.

Physically, computer memory is and remains simply a vector, a one-dimensional series of cells with ascending addresses. This one-dimensional structure is the simplest possible, and indeed the programmer may design his logical space in the same linear fashion. If his data is a list of numbers to be added in their original order, then the space may reasonably remain one-dimensional, with one number assigned to each of a series of contiguous data locations. Even here, the programmer thinks in spatial terms: each number occupies a location perhaps two bytes in length, so that a list of ten thousand numbers would occupy a region of twenty thousand bytes. The programmer must request that much space from the machine.

More generally, the structure is made to fit the problem. If the problem is mathematical and requires a matrix (for example, three rows and three columns of numbers), then clearly the elements should not be arranged in a simple vector. The numbers in each row have a special relationship, and so do those in each column. Programming languages provide the needed structure in a data type called an array, which can be two-dimensional to reflect the rows and columns. The logical space now differs from the physical storage, which remains a vector; the programming language takes care of the difference (figure 6-1).

Mathematical data are not the only data that lend themselves to such constructions. Indeed, mathematical structures tend to be simple, either vectors or multidimensional arrays. The logical relationships among letters, words, or arbitrary symbols are often more complex than numerical relations. The most common data structure beyond the array is the *tree*. Computer specialists, alas, seldom bother about consistency in their metaphors; their trees have branches as we would expect, but the meeting point of two branches is called a *node*, and they further distinguish nodes as parents and offspring (or fathers and sons, mothers and daughters). A tree is two-dimensional: its nodes contain data, and its branches are links from one node to another, directed links that point from the father node to the son. True to the biological comparison, a node may have many offspring but only one parent. A *leaf* is the end of a branch, a node with no offspring. One detail does not conform to the metaphor at all: the root, the oldest ancestor, is generally placed at the top of the diagram. A computer tree grows downward.

All of this seems highly artificial, logical game playing in the tradition of Turing. Yet the tree is a remarkably useful way of representing logical relations in spatial terms. The tree structure itself carries information, as we can see from the following arithmetic example. The expression $3 + 4 - 3 - 2 \times 6$ is at best unclear and at worst ambiguous until we establish an order of precedence for our arithmetic. But a tree structure makes the expression transparent (figure 6-2). Every number becomes a leaf, every operation an intermediate node. The rule is: to evaluate the tree, begin at the leaves and apply the parent operation to each pair of numbers; then replace the parent node by the resulting number and continue. Compute $3 + 4 = 7$, then $7 - 3 = 4$, and so on. The tree tells us that the expression should have the following parentheses: $[(3 + 4) - 3] - (2 \times 6)$. The tree itself conveys the order of operations, the logical structure of the expression.

Figure 6-1. Array Representation

87
Electronic
Space

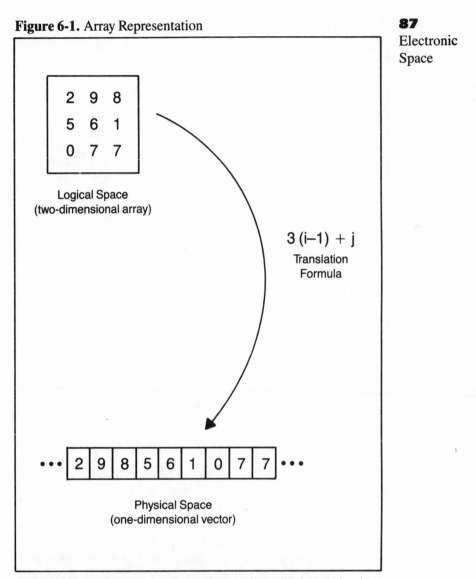

Logical Space
(two-dimensional array)

3 (i–1) + j

Translation
Formula

Physical Space
(one-dimensional vector)

The physical address space of a computer is a one-dimensional vector, but the logical space may have as many dimensions as the programmer cares to imagine. He envisions, for example, a two-dimensional array (with three rows and three columns). In his programs he uses ordered pairs (i, j) to refer to elements of the array. Thus, (2, 1) refers to the element in the first column of the second row, or 5. The computer actually stores the array as a vector of nine numbers. It uses the formula 3 (i−1) + j to convert the ordered pair (2, 1) into the proper location —3 (2−1) + 1 = 4—and fetches the fourth element in its vector. Such formulaic manipulations (invisible to the programmer) give computer space its strict mathematical and geometric quality.

Figure 6-2. Tree Diagram for Arithmetic Expression

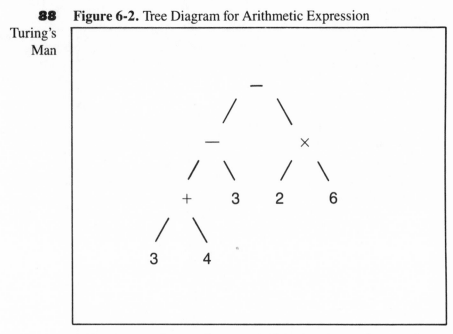

Arithmetic expressions are often converted into a tree structure for processing.
Above is the tree structure for [(3 + 4) – 3] – (2 × 6).

Programmers build trees in order to capture such structures for
simple tasks like arithmetic and for more complex symbolic manipulations as well. A list of names can be sorted by building a
tree structure; a bibliography can be read into a tree structure and
then queried; even chess programs build trees to evaluate lines of
play. The beginning of a decision tree for the opening moves in
chess might resemble the one in figure 6-3. The first node belongs to white, the sons of this node to black, the grandsons again
to white, and so on. The program exploits the tree by following
branches until it comes to a leaf that is marked as a favorable or a
poor position. Chess programs often store hundreds or thousands
of book openings in this arboreal fashion.

There is a vast literature on the properties of trees and other
similar structures with such colorful names as doubly linked lists,
last-in-first-out stacks, and heaps. All serve to relate elements of
data in the computer's logical space; all depend somehow on links
and nodes, on storing information in cells and establishing pointers among these cells. The result is a geometric pattern that reflects the information contained. Such diagrams are nothing new.
Genealogical trees have been used for centuries. For a hundred

Figure 6-3. Tree Diagram for Chess Moves

89
Electronic
Space

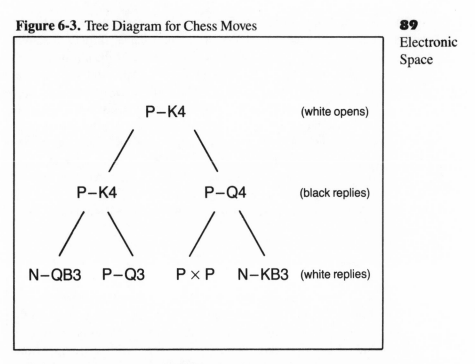

A decision tree for some opening moves in chess.

years, scholars have drawn similar diagrams to illustrate the lineage of medieval manuscripts. But husbanding manuscript trees was an esoteric art. No one spoke of manuscript or genealogical space because there was no physical embodiment for these trees. Electronic trees, however, are embodied in the circuits of computer memory, and there they have become the ubiquitous tools of millions of programmers.

Electronic trees are not as rigid as genealogical trees, which grow only at the leaves and not according to the genealogist's wishes. The programmer makes his trees and modifies them at will, inverting or pruning off whole branches to suit his problem. The tree that represents possible openings in chess need not be a frozen structure. The program may alter the tree to reflect its own success or failure. It may add a new branch by recording a move not in the book, a move that led to a win in a previous game. If it consistently loses by using one line of play, it may decide that the whole branch is not worth saving and prune it to make room (always a precious commodity in a chess program) for another line from the book.

In short, a program's logical space is alive and capable of endless variation, either by decision of the programmer or automatically in the course of the program. Like an artist's medium, computer space has structure and plasticity. At the physical level, memory remains a series of numbered cells, and the cells, of course, never move, but the numbers allow them to be reordered logically into arrays, trees, or any other hierarchy the user can imagine. In representing anything from numbers for arithmetic to automobiles for traffic simulation to pieces on a chess board, the programmer is master of a powerful logical geometry that seems to reflect the widest variety of problems that arise in the real world—a geometry far more down-to-earth, therefore, than the admittedly more precise sciences of Euclid or Riemann.

Finite Place

Space is what each culture chooses to make of it, and what each chooses depends upon the tools and techniques available. To the ancients, space was felt in the press of clay on a potter's wheel, studied by bisecting angles in the sand with a compass and straightedge, or thought out while strolling through the court of the Lyceum. To Western Europe, space was occupied by rarefied or condensed steam in an engine, marked out as a grid to represent equations in analytic geometry, or hotly debated at meetings of the various national academies. Computer space resembles in some ways the mechanical-dynamic space of Western Europe, in others the manual space of the ancients. What is new is not any one element but rather the particular combination of elements from the past.

Let us begin with the more remote culture. The ideal way to appreciate the ancient sense of space would be to visit Athens in the fifth century—to walk through the full and chaotic marketplace and then ascend to the Acropolis, whose buildings stand out assertively from the rock, each declaring its geometric perfection and plenitude against the skyline. But with the Acropolis in ruins as it is now and the life and craftsmen of the agora gone, the feeling is hard to capture. We can now catch our best glimpse of ancient space from literary monuments, from the philosophy and mathematics that have survived more or less intact. These, together with the physical remains, teach us that one spatial idea—the principle of the immediate and the finite—underlay all ancient technology and philosophy, as well as the magnificent theorems of geometry.

The importance of those theorems is indeed great. The synthetic geometry eventually codified by Euclid was no mere mathematical construct to the ancient mind; it was an explanation of the world. When the Pythagoreans claimed that everything was number, they were thinking of geometrically realized numbers that physically composed the universe. They did not define their numbers in the abstract terms of modern set theory but instead attributed shapes to numbers: plane geometrical shapes. Following the Pythagoreans, Plato derived all matter from four elements of ideal polygonal shape (pyramids, cubes, octa- and icosahedrons). Since these polygons were composed of right triangles, all of space in Plato's universe was in fact made up of the simplest rectilinear figure. The reduction of matter to geometry, the desire to claim a fundamental order for things, the preference for straight-line figures—all these qualities were characteristic of Greek cosmological thinking as well.

Aristotle's physics, perhaps more influential than any other, consciously rejected the excesses of Pythagoreanism. Nonetheless, space in his philosophy remained concrete and palpable, like Platonic geometric space, and never became the abstract coordinate system it later became in modern physics. Aristotle in fact preferred to speak of "place," which he defined as the "limit of the surrounding body." Empty space had for him no reality. Indeterminate matter created a place for itself, filling space and forming the basis of all objects in the world. When Aristotle looked at a thing, he saw informed matter; the surrounding space had no significance, as it was really the place being occupied by some other informed matter. For the Aristotelians, there was no such thing as a vacuum. The universe was a plenum, space was the sum of all the matter in the world, and the sum of matter was finite—these views were shared by most Greek philosophers except the atomists. Aristotle's and Plato's world was neatly limited by the final sphere of the prevailing astronomy. Beyond that there was nothing, neither body nor place. As for space in the more modern, Newtonian sense, there was not room for that either outside or within the boundary of the crushingly full world of Aristotelian physics.

What of those troublesome atomists such as Leucippus and Democritus? They were the only philosophical sect to assert the existence of the void, of empty space. The universe, they claimed, was made of tiny atoms colliding in an endless space, atoms whose random collisions eventually accounted for all the organized structures and creatures in the world. Such ideas shocked the ancients precisely because of the tendency to view the cosmos as a full and orderly system, and it is appropriate that atomism was taken over as the official physics of the Epicureans, who specialized in scandalizing the moral and religious beliefs of the ancient world as well. But even the atomists did not really abandon the Greek love for plenitude; they simply shifted it down to the scale of the atoms themselves. Even for the atomists, space did not fully exist; it was instead the field in which atoms could act and react. These atoms were solid, indivisible units of matter: they were all the reality there was, composing everything we see, know about, or can know about. As Oswald Spengler realized, the modern notion of atoms as tiny fields of force (mostly empty themselves and inhabited by such phantom entities as electrons and protons) bears faint resemblance and owes little intellectual debt to ancient atomism. Like the other ancient philosophers, atomists too sought to preserve the notion of plenitude, of the fullness of existence, in the nuggets of reality called "atoms" or "indivisible bodies."

One way or another, the ancient world was philosophically a full place. It was a perfect home for the Greek craftsman and artist, who may never have read Plato or Aristotle but did put their philosophies of space into technological practice. In the Greek crafts, a love of line and of plastic, tactile forms is found on every scale. Greek vases declared that the world was a plenum when they defined space by occupying it in such a masterful way. The painted figures on vases (during the best periods, before about 450 B.C.) were little more than silhouettes and were grouped in ways that highlighted the shape of the pot. Meanwhile, the Doric temple and all its variations made the same assertion of occupied space on a grand scale. The temple was not built to blend into the scenery or to soar endlessly upward like a Gothic cathedral, but rather to fill space with an ideal geometric structure and so to assert its independence from its surroundings. Siegfried Giedion understood the analogy to Greek cosmology when he wrote: "As the Greek temple symbolizes forces in equilibrium, in which neither verticals nor horizontals dominate, the earth in the classical view formed the forever immovable center of the cosmos" (*Mechanization Takes Command*, 15). The Greek temple oc-

cupied and asserted its place with no less authority than the earth
itself in Aristotle's geocentric physics. We can sense this geometrical balance of forces even in the ruins of Greek architecture
that remain; the buildings of the Acropolis are in one sense giant
marble illustrations of Euclidean geometry.

Greek temples, statues, pots, and even chairs are examples of
crafted space. All reveal the craftsman's hand informing space by
molding the material that will occupy it. To speak at all of the
"use of space" in art and architecture is to suggest the artistic
values of a later, distinctly nonclassical era, for a Greek artisan
manipulated matter, not space. For the potter at his wheel shaping the clay, the emphasis was upon the tactile, the concrete, and
particularly the linear. And these were generally the values of the
"classical" trend in art, whenever it returned in the Western
tradition.

Infinite Space

Western Europe found it easier to modify ancient architectural
and artistic uses of space than to break through to new philosophical and scientific notions. Architects were building Gothic cathedrals while philosophers were still rehashing Aristotle's physics.
From the Renaissance on, however, the trend in science too was
away from the plenum of the ancients toward the idea of empty,
bodiless space, the idea of pure extension. Of great importance
was the perfection of analytic geometry by Descartes and others.
These mathematicians defined a new relationship between number and geometry, which had previously been the only science of
space. Points on a line or in two or three dimensions could now
be put into correspondence with algebraic (and eventually real)
numbers, and equations could now describe curves in a plane or
in space. Mathematical space became a coordinate system, the
xyz-axes known to us from high school algebra. In the Euclidean
view, it had been the ideal receptacle of plane and solid geometrical shapes, constructed not according to an equation but by the
(craftsmanlike) use of compass and straightedge. Eventually, it
was seen that Cartesian space was continuous or complete, that
lines in space were composed of an infinite set of dimensionless
points, each of which corresponded to one real number.

The solidity of ancient space then dissolved. Points were now
defined by the algebraic relationships of their coordinates: they

existed in space, whereas the ancients had thought just the reverse, defining space by virtue of the figures that occupied it. Spengler captured perfectly the contrast between ancient and later Western mathematics: "The beginning and end of the Classical mathematic is consideration of the properties of individual bodies and their boundary surfaces; thus indirectly taking in conic sections and higher curves. *We*, on the other hand, at bottom know only the abstract space-element of the point, which can neither be seen, nor measured, nor yet named, but represents simply a centre of reference. . . . The Classical mathematician knows only what he sees and grasps. . . . The Western mathematician, as soon as he has quite shaken off the trammels of Classical prejudice, goes off into a wholly abstract region of infinite numerous 'manifolds' of n (no longer 3) dimensions" (*Decline of the West*, 1:81–82).

This new mathematical space was understood in intimate connection with a new physics by such men as Gassendi and Newton. When Newton explained the course of the planets in mathematical terms, his equations were those of analytic geometry applied to objects in the physical world. Physics had become applied mathematics, in contrast to the nonmathematical speculations of the Aristotelians. And if the planets followed elliptical curves described by the equations of gravity and the inverse-square rule, it was natural to think of the physical space through which they traveled as a kind of Cartesian coordinate system. Newton himself wrote: "Absolute space, in its own nature, without relation to anything external, remains always similar and immovable. Relative space is some movable dimension or measure of the absolute spaces; which our senses determine by its position to bodies" (Scholium to Definition 8, *Sir Isaac Newton's Mathematical Principles*, 1:6). Newton's absolute space was the ultimate reference, which allowed the physicist to establish coordinate systems in relative space for measuring and relating the various movements of the planets and stars. Absolute space guaranteed the mathematization of space that was central to Newtonian physics.

For a time in the seventeenth and eighteenth centuries, there was a violent debate over the fullness of space. Such mechanists as Huygens, Leibniz, and Descartes himself agreed on logical grounds that there could be no space without matter; Aristotle had proven the same thing two thousand years before. In order to eliminate the logically odious notion of action-at-a-distance, Des-

cartes and Huygens evolved elaborate mechanisms to explain
even gravity as the interactions of vortices of matter in immediate contact. Newton, less a philosopher in the old sense and so a better scientist in the modern sense, realized that the idea of empty space met more reasonably the requirement of the new physics. To fill space with an unnecessary sea of matter was merely to clutter up the beautiful, austere equations. Gradually, all scientists came around to this view and began to grant empty space the same status as matter itself.

Newton was so excited by the qualities of this static coordinate system, this framework for all physical laws, that he spoke of absolute space in divine terms. "[God] endures forever, and is everywhere present," he wrote in the *Principia*, "and by existing always and everywhere, he constitutes duration and space" (Scholium to Book 3, *Sir Isaac Newton's Mathematical Principles*, 2:545). This hint of pantheism brought Newton's follower Clarke into conflict with Leibniz, and their debate remains the best example of the emotions aroused by speculation on the nature of physical space. Newton was proved right, frighteningly right, early in this century, when physicists learned that the densest ordinary materials are almost entirely empty space, in which tiny subatomic particles wander about. Indeed, if the universe were a plenum, as Aristotle and even the forward-looking Descartes supposed, it would apparently collapse to a single point under its own mass.

All this was heady stuff: mathematical physicists of the age were debating issues that in the ancient world had belonged exclusively to philosophers. Physicists in Newton's day were rightly called natural philosophers because they brought together abstract speculation and observations from nature into a new synthesis. Proof came when mechanical and dynamic technology could actually build machines to operate on scientific principles in Newtonian space. The finest achievement was the steam engine. Pascal had already shown that variations in air pressure only made sense if the vacuum existed; von Guericke had demonstrated the force that a vacuum could exert when a team of horses failed to separate two halves of an evacuated sphere that were held together by nothing but the push of the atmosphere. In the eighteenth century, the steam engine realized the promise of von Guericke's experiment. Empty space not only existed but could be put to work. Indeed, steam and other atmospheric engines exploited space itself as a kind of fuel to produce controlled power.

These new engines established relationships of mechanical advantage between parts of space; some of the principles here were as old as the winch and the lever, but they were applied in the Industrial Revolution with a new enthusiasm. The engines also exploited the heat advantage between regions of space, a technique practically unknown to the ancient world. The ancient potter or architect filled up space with matter to create functional and pleasing shapes, but the designer of the steam engine evacuated matter from space (in the piston) to make his machine perform useful work.

At least one purely artistic trend, by the way, reflected the new attitude toward space. Baroque artists, working in the same century as Galileo, Descartes, and Newton, began to dissolve the rigid lines and plastic forms that characterized Renaissance painting and architecture, forms that Renaissance artists had derived in part from their classical models. The emphasis was now upon the visual rather than the tactile, upon the drama of light and space rather than the pleasing geometry of line. Altogether, it was an appropriate art for the century that discovered the scientific nature of space and began to measure and use atmospheric pressure.

The Geometry of Electronic Space

In one sense, there has been a simple development in the concept of space from ancient to Western European to contemporary culture, a growing awareness of the possibility of empty space as distinct from matter and a desire to manipulate this empty, mathematical space. Greek craftsmen manipulated matter, not space. The Parthenon is no less a masterpiece than the cathedral at Amiens, but the enclosed and soaring space of the interior of the cathedral is very different from the occupied and carefully delimited space of the Greek temple. The technology of Western Europe manipulated matter in space (through rarefaction and condensation), still not space itself. Newtonian space remained absolute and unalterable, the rigid framework by which all change could be measured.

It is the mathematicians and physicists of our century who have put forth the idea that space itself can be manipulated, subject to geometrical changes. As usual, the mathematicians have led the way; already in the nineteenth century, they had discovered geometries that denied the intuitive notion of "flat"

space sanctioned by the Euclidean tradition. In the twentieth century, the esoteric science of topology has found an endless number of spaces with different properties; mathematicians began to apply the term "space" metaphorically to any collection of objects that satisfied certain postulates. They began to create spaces, to construct, explore, and manipulate these collections of objects for their own curious purposes. Their extraordinary abstract reasoning was applied to physics by Einstein and others, who argued that our universe is in fact a non-Euclidean, "curved" space, replacing Newton's absolute coordinate system with a system whose very geometry is distorted by the presence of matter, which manipulates the space around it through its gravitational force.

Computer space is another territory opened up by twentieth-century discoveries. In previous ages, there has been a division between the technology of space and the philosophy of space. Both reflected the same underlying cultural value—the finite space of the ancients, the infinite space of the later European thought—but the craftsman and the thinker were usually very different people who took wholly different approaches. One saw his work realized in stone, clay, or metal; the other sat down and thought, and perhaps described what he thought in prose. The mathematician has always been the most craftsmanlike of philosophers; the spaces he creates (geometrical and now topological) do seem to have a continued existence and can be examined and even modified by others. Even more than the pure mathematician, Turing's man is both a philosopher and a craftsman of space: he treats space in the machine as a plastic, almost palpable material and yet remains a logician who calls forth data structures from his imagination. Because computer space is both physical and logical, Turing's man is therefore able to absorb elements from all the past traditions of craft and philosophy.

He is in tune with the most recent science and mathematics. Even the cosmology of Einstein bears a relation to the electronic concept of space. Aristotle's plenum accorded nicely with the practice of the Greek craftsman; Newton's vacuous space and mechanical universe corresponded to the clock and steam-engine technology of his culture; and in the same way, Einstein's curved space is the scientific counterpart of the programmer's electronic workspace. The physicist has taught us that space itself can be manipulated. Although admittedly we do not have the technical resources to put the manipulation of physical space to work for us, we are accustomed to the idea that gravity is geometry, that matter bends space. It is natural, then, to use the metaphor of manipulated space where it does make practical sense—in the new electronic technology. The computer programmer lives in two worlds simultaneously: his space is a logical entity (like the space of the topologist), and yet it is logic realized in the transistors of storage. In this way, too, it resembles Einstein's space, which is mathematical and abstract and still claims to be the space of the world of experience.

Electronic space is much easier to understand than its counterparts in mathematics and physics. It is, after all, an artifact, a constructed space that must function in thoroughly predictable ways in order to serve its technological purpose. Artifacts are almost always less sophisticated than the leading researches of sciences because they must be built upon ground that is already intellectually secure, not at the frontiers. A computer memory is mathematically quite simple, a coordinate system of the type pioneered by Descartes, and computer space is, like Newtonian space, a straightforward application of number to geometry. Computer space also has properties that take us back to the even more intuitive notions of the ancients. Inside the machine, numbers are as discrete and corporeal as even a Pythagorean would have them. Moreover, computer space is always finite, like the universe in most ancient cosmologies. And the emphasis in finite space is on compactness and parsimony. In a small room, more care must be taken to assign everything its place; so in a small workspace logical entities must be carefully crafted to fit together and to waste as little space as possible. Arrays must not exceed the size dictated by the data, and trees must be pruned of unneeded branches. In sum, electronic space has the feel of ancient geometric space, but it has been mathematized and numbered according to Cartesian principles. Finally, it has derived its logical foundations from the twentieth-century revolutions in science.

What is unique about electronic space, unique in the history of technology and the philosophy of space, is its popularity, its accessibility. Computer flowcharts, tree structures, and lists have invaded every discipline. Our age in general is mad for diagrams; natural and social processes and even human thought are commonly pictured in spatial terms by psychologists, linguists, anthropologists, sociologists, and social planners, as well as physicists and chemists. Using treelike designs to describe the creation of sentences came into fashion with Chomsky and his followers, in close connection, by the way, with the development of high-level programming languages. Today, anthropologists and even literary critics are diagramming everything from social relationships in aboriginal tribes and primitive myths to urbane modern poetry.

Drawing a diagram is nothing new, but the kinds of experience that are viewed in these terms are surely as great now as ever before. The computer with its data structures is at least partly responsible for this trend. For the computer makes a geometer out of anyone who has a computational problem to solve. It requires the user to interpret his problem in spatial terms, to consider the amount of storage his program will need and the logical structure he will give his region of storage. The finite, discrete geometry of the computer extends itself over vast areas of intellectual activity, areas too imprecise for the geometric systems of contemporary mathematicians.

7 Time and Progress in the Computer Age

There is an intimate connection between a culture's attitude toward time and the technology by which it measures time. Which is cause and which effect are not easy to say. Did the Europeans invent weight-driven clocks because of their desire to know the precise hour, or did the clocks, once invented, change their way of scheduling business and social activities? The answer is likely to be yes to both questions. Some cultural value must have driven the Europeans to perfect a device that no other culture had cared much about. Yet once the new technology was called forth, it proceeded with its own relentless logic and eventually helped to reorder the values of the whole culture.

Did we invent computers because we needed very fast calculators, or did the calculators suggest to us the importance of solving problems that require such speed? In either case, our appreciation and our evaluation of the passage of time is changing in the computer age. Logical events occur in a computer with a swiftness that we cannot compare to anything in our daily experience. Much of the efficacy, and some of the charm, of the computer lies in its sheer speed. There are endless anecdotes: the computer can solve in x seconds a problem that would require y mathematicians z centuries to complete. There is also a serious point to be made: anyone who works with the computer is introduced to a second temporal world, a world of nanoseconds (billionths of a second). If at first we try to understand computer time in terms of previous experience, our outlook is soon reversed.

The way the computer "processes" this commodity becomes a
model for our thinking about the passage of time in nature and culture in general.

There is another side to the issue. Men, women, and the cultures they create all exist in time, their work proceeds through time, and the passage of time marks out success or failure. A culture's optimism or pessimism about the future is also in part determined by its technology. The perception that things are getting better, remaining static, or degenerating rests heavily upon the ability or inability to push things in the right direction, particularly in the material sense—more food, better shelter, more comfort. The Greek pessimism must have been due in part to the failure steadily to improve the control of nature; the optimism that reigned for centuries in Western Europe coincided with the ability to harness natural forces more effectively. The belief in progress is simply faith in the ability of mankind to work through time; time is no longer an enemy causing only human decay but an ally for the betterment of the human race.

The effect of the electronic technology upon the notion of progress is fascinatingly complex, for it manages to suggest both Greek pessimism and Western optimism. The synthesis comes from a new intimacy with time, as both ally and enemy: previously men and women lived in time and worked through time, but Turing's man is the first who actually works *with* time. Like space, time is a commodity provided by the computer, a material to be molded, insofar as this is possible, to human ends. This intimate contact with time promises success in time (progress) but also an awareness of ultimate temporal limitations.

Electronic Clocks

The clock has been at the center of Western technology since its invention in the Middle Ages. Computer technology too finds it indispensable, although it has changed the clock from a mechanical device to a wholly electronic one. Every digital computer has one or many timers built into its central processor. The timer is often a device that releases electrical charge at intervals, providing a regular pulse for the machine's operations; the interval may be measured in nanoseconds. For the computer, the passage of time is not a continuous flow, as it often is in human experience: time passes as a series of discrete units, which mark off the prog-

ress of the machine. The purpose of these units is to allow the
central processor to execute one by one the instructions given in
its program. In the computer, as elsewhere, no physical process is
instantaneous. Since the processor uses electric current to carry
out its calculations, in the end all computation is simply the
movement of electrons. An electronic signal propagates through
a wire at an enormous speed, yet even this speed is finite: it takes
a measurable amount of time for a signal to travel even from one
point to another on the same silicon chip. More time is required
for the signal to pass through complicated circuits that perform
logical operations. The electronic timer provides the measure by
which the processor ticks its way through its calculations, ensur-
ing that the electrons have settled down, that one step is finished
before the next is begun. The instructions themselves may require varying amounts of time; an add instruction may need only
a fraction of that required by a more complicated division instruction. This variation must be taken into account by the sequencing mechanism, which decides how many pulses of time to
allot to each instruction. Even for a slower minicomputer, it is a
question of perhaps one hundred millionths of a second instead of
ten millionths, and the reader may well ask whether these tiny
fractions of time saved or lost really matter.

In fact, an engineer or programmer is no happier with a CPU
that wastes time then with a motor that squanders gasoline, for
time is what computers were built to save. The CPU (and each
program it executes) is a consumer of time, of the abstract, electrical pulses handed out by the sequencing mechanism; it therefore must be constructed to be a wise consumer. Time is a resource, perhaps the primary resource, by which the computer
operates, like the water that powers a mill or the coal in a steam
engine. Electronic clocks are the computer's way of measuring
out the electricity that is the ultimate source of all its functions.
The fact that these clocks serve such varied purposes as sequencing, calculation, and the storage and retrieval of data is another
indication of the remarkable unity of the von Neumann machine.

In one sense, the computer processes binary symbols stored in
an electronic medium, and the clock measures its progress. In another sense the computer processes time itself, by transforming
billions of contentless pulses of electrical energy into useful instructions for manipulating data. An ordinary clock produces
only a series of identical seconds, minutes, and hours: a com-

puter transforms seconds or microseconds or nanoseconds into information.

The enormous speed of this transformation puts the computer's operation in a temporal world that is outside of human experience. A program is a recipe, a list of commands, each of which can be executed in a fraction of a second. A programmer can read the commands, alter them, enter them in machine-readable form; but once the program is loaded into core memory, the programmer can have no feeling for the course of its execution. He may be able to watch a tape containing his data spin or hear the clicks of the reading mechanism as it flies over the surface of a magnetic disk, but usually the computer installation is serving a number of programs at any one time (*time-sharing*). The programmer does not even know the exact moment that the CPU is devoting to his program. Events in a computer are simply beyond the range of human senses. Just as electrons are too small to be seen, so their movements are too fast to be appreciated. At any rate, the course of an individual electron does not tell us how a program is doing because the logic of computers is expressed in forces that are averages of the behavior of many electrons. No machine has ever been so far removed from the world of human experience: the largest aircraft carriers are still infinitely closer to the human scale than the simplest, slowest microcomputers.

Engineers who design computers do need to know what is happening in the CPU at each nanosecond, for their job is to arrange and rearrange the signals that perform the sequencing and logic. They, more than the programmer, must find some way into the hidden temporal world of the machine, and so they build electronic prostheses, oscilloscopes and the like, that usually turn split-second changes in the circuits into patterns on a television screen. They can sense more than anyone the paradox of this second temporal world. Tracy Kidder, who has written on the psychology of these engineers, quotes one: "I feel very comfortable talking in nanoseconds. I sit at one of these analyzers and nanoseconds are *wide*. I mean you can see them go by. 'Jesus,' I say, 'that signal takes twelve nanoseconds to get from there to there.' Those are real big things to me when I'm building a computer. Yet when I think about it, how much longer it takes to snap your fingers, I've lost track of what a nanosecond really means" (*Soul of a New Machine*, 137). If it takes, by the way, half a second to snap one's fingers, that is five hundred million nanoseconds.

This radical separation between the human scale and the scale of one's work is not an experience limited to the computer specialist. Electronics is perhaps the most prosaic field in which the separation occurs. Astronomers and nuclear physicists were familiar with the problem of scale years (in the case of astronomy, centuries) before the first computer was built. Astronomers routinely speak of processes such as the creation of a star requiring millions of years or the development of a galaxy requiring billions. At the same time, they deal with speeds faster than those of electrons in a circuit: the speed of light and other electromagnetic radiation in a vacuum. And they describe commensurately enormous masses and volumes of space. Any attempt to depict these numbers graphically, to find an analogy that would bring them more meaningfully into human comprehension, must take on the melodramatic quality of science fiction. Scientists have developed a special notation that allows them to work mathematically with vast distances, masses, and durations, but such a notation does not bring these enormous figures any closer to human experience. Nuclear physicists have the same problem in the other direction because the particles or entities they work with are smaller, faster, and even more ephemeral than the electrons of the computer. They can detect these entities and measure their properties only by the most delicate and tentative of experiments involving particle accelerators, electronic counting devices, and indeed computers. Beyond all this equipment lie sophisticated mathematical tools.

Like the computer specialist, the physical scientist must use indirect methods to detect changes that are too slow, too fast, too large or small for human perception. In that respect, they are simply carrying to extremes the principle established in science since the invention of the telescope and microscope, a principle so firmly established that it can now serve engineers as well. This distinction must be kept in mind. Physicists deal with minute quantities because they want to study the constituents of atoms, and atoms happen to be very small. As scientists, they are pursuing nature where it leads them. Computer specialists, like other engineers, are not directly engaged in the study of nature, and it is in no way accidental that their programs execute so quickly. Computers are designed to do what they do; they constitute one of the more remarkable human manipulations of the natural world. The whole point of creating microscopic artifacts by electronic engineering is to take advantage of our knowledge of solid-state physics to invade the fast and tiny world of electrical phenomena.

Unlike his colleagues in physics or astronomy, the computer specialist *chooses* to work with phenomena on a wholly different time scale from his own: he is an artificer rather than a scientist. And as a manipulator rather than an observer of microscopic time, his concern is not simply to understand the world of solid-state physics inside his machine but to structure and control that world. Of course, the pioneering work of the physicist must precede the manipulative work of the engineer because he can only build electronic components when the principles have been explained, at least to some extent, and modeled mathematically. The physicists are always ahead of the engineers, exploring particles and events on an ever smaller scale. It will be a long time, surely, before quarks serve as components in a computer.

The attitude of the pioneer differs from that of the colonist: the computer specialist has more pragmatic, perhaps less imaginative goals, than the physicist. Yet his prosaic point of view makes him all the more aware of the contrast between human time and time in the computer. The physicist may live out his entire intellectual life in the world of subatomic particles, but the computer specialist must constantly return to the tortuously slow pace of human activity, simply because he wants his machine to perform work that men and women can appreciate and use. He must reckon not only with the nanoseconds of CPU time but also with the minutes that programmers have to wait to receive their answers. He understands that the computer demands a mathematical precision in the measurement of time and an abstraction from human experience beyond any other machine ever made.

Time Experienced and Measured

The Greeks measured time with the same devices available to most developed cultures lacking mechanical technology. They had sundials and devices for charting the sun's course along the ecliptic, its apparent path through the heavens. Like the Egyptians and Babylonians, they were able to make fairly accurate calendars. For shorter periods of time, they used the *clepsydra*, the water clock. But knowing the exact passage of an hour was important only in special circumstances: the clepsydra was used in the law courts to limit a speaker's flights of rhetoric. The everyday measurement of time was by our standards extremely subjective. The ancients rose early, more or less with the sun, and counted their day by hours of sunlight, which their dials would tell them or which they could reckon themselves by the sun's position. For later Greeks and for the Romans, an hour was one-twelfth of the day or the night, and so it varied seasonally.

The absence of ubiquitous, accurate clocks must have made their perception of the day very different from ours, and if we could travel through time and spend six months in ancient Athens, we would find that perception a major barrier between ourselves and the inhabitants. No appointments to be kept to the minute, perhaps even to the half hour, no means of knowing exactly when dinner was served, and so on—it would be quite disorienting for a modern man or woman. Timekeeping is a tyrant whose rule has been with us for so long that we would be frightened by its absence. The ancients measured time by their experiences. The hours a Roman elegist spends with his mistress seem but a few minutes; the minutes Horace spends with a bore are endless. Even the regimented day of a Greek religious festival or a Roman day in court must have possessed a spontaneity, a flow determined by events rather than by the clock. We surely spend far more time thinking about time than the Greeks or the more pragmatic Romans.

The Greeks did reckon with larger units of time. They used calendars just as we do, and they knew of course that time imposed upon men the ultimate limit of mortality. They did not generally see time as something to be used: "time is money" is a cliché not to be found in Greek literature. Nor was time so important a measure of accomplishment. A craftsman may have had to work toward a deadline, but his tools were not measured by their output per unit of time (a characteristic development of mechanical technology). In any case, Greek machines were immediately dependent upon the men who used them, so that the number of pots a wheel could produce depended largely upon who was doing the throwing, although a wheel that turned more easily and evenly would increase productivity. It is pure speculation to try to reconstruct the thinking of an Athenian potter because Greek craftsmen are known to us only through their artifacts or through the biased accounts of well-to-do and often aristocratic philosophers and orators. Still, it is hard to imagine an Athenian factory owner devoting much thought to increasing his profit through greater productivity—more pots per potter per day. As M. I.

Finley points out, "Technical progress, economic growth, productivity, and even efficiency have not been significant goals since the beginning of time" (*The Ancient Economy*, 147). They were not goals of the ancient craftsmen. Such concepts demand a notion of time that was first developed by men of the Middle Ages and later, namely, time as a measure of economic progress.

The birth of that new notion was celebrated by the bells of the mechanical clock, invented around 1300, which freed men's ability to tell time from the circumstances of the weather. The sundial could not be consulted on cloudy days, and water clocks were cumbersome and subject to freezing in the climates of northern Europe, but the verge and foliot clocks depended upon gravity, which never failed them. So began the abstraction and mathematization of time, which has continued to the present day. When the pendulum was introduced in the seventeenth century, time could be divided into fine units, minutes and even seconds. Men and women could organize their day around these arbitrary divisions. Thus, "abstract time became the new medium of existence. Organic functions themselves were regulated by it: one ate, not upon feeling hungry, but when prompted by the clock: one slept, not when one was tired, but when the clock sanctioned it" (Mumford, *Technics and Civilization*, 17).

Somewhere along the way, then, perceptions were reversed: men and women ceased thinking of a minute as a length of their experience and started measuring their experiences by minutes of the clock. The measurement assumed ever greater importance in the eighteenth and nineteenth centuries, with the invention and perfection of the steam engine and prime movers. Men used mathematical time to regulate their new machines, which worked more or less independent of human operators. Soon after Watt built his first improved steam engines, he began charting their efficiency in terms of work done per unit of time—horsepower, as he called it in an effort to explain his achievement in familiar terms—and sold machines on the promise of this performance. In the nineteenth century, the mathematization of time, which had always been a matter of abstracting away from nature and human experience, returned to conquer nature, in the form of steam and gasoline engines and the dynamo.

The mathematical philosophers, such as Descartes and Newton, had long before realized the significance of this new kind of time and had made it the basis of physics, the first modern science. They thought of all motion as a function of time, measured in arbitrary, identical units. This mode of thought accorded nicely with the concept of inertia: that constant rectilinear motion does not require an explanation any more than a state of rest does, but a change in motion, an acceleration, must have a cause. To understand acceleration mathematically as a changing rate of change, requires a sophisticated notion of time, but once this is achieved, the scientific gains are tremendous. A constant force accelerates a mass; acceleration multiplied by time equals velocity; velocity multiplied by time equals displacement; and so on. These principles today are easily within the grasp of high school students, yet it required centuries of effort (from Buridan and Oresme in the late Middle Ages to Galileo, Descartes, and Newton in the seventeenth century) to get them straight. Today in physics, the variable t (for time) is ubiquitous, and it has been found that most natural phenomena that yield to mathematical description at all can be described as differential equations in time—changing rates of change. Newton himself expressed this new way of looking at the flow of natural events when he postulated absolute time, along with absolute space, as a fundamental quality of the universe. "Absolute, true and mathematical time, of itself, and from its own nature, flows equably without regard to anything external," Newton wrote in the *Principia* (Scholium to Definition 8, *Sir Isaac Newton's Mathematical Principles*, 1:6). Our senses tell us, sometimes accurately, sometimes not, of the passage of this true time; but our sense perception must clearly take a second place. The abstraction is the reality. In Newton's equations, real time goes on forever.

All machines, and particularly the gear-driven mechanisms of the Industrial Revolution, are inadequate to represent Newton's "equably flowing" abstract time. All must suffer some inaccuracy. Yet the increasing precision in mechanical clocks, engines, dynamos, and scientific instruments throughout the last two centuries has shown how important the ideal has seemed to Western man. Computer time is the acme of this development, the most abstract and mathematical notion of time ever incorporated into a machine; it takes the scale of time far beyond the lower limit of human perception. It represents the final triumph of the Western European view, in which time itself becomes a commodity, a resource to be worked, much as a structural engineer works with steel or aluminum. Like these metals, computer time is a finite resource, and the computer programmer is always concerned with using it as sparingly as possible.

Already a century ago, time had been mathematized by scientists and made into a measure of productivity by industrialists. In pre-electronic technology, however, it remained always an extrinsic measure of a product. A steam turbine's performance was measured by revolutions per minute, but its product was the work it accomplished through those revolutions. Although it is common today to speak of time as a commodity, in most cases it is not so. A lawyer may say that he is selling you half an hour of his time, but if he really spent that half hour reading off the minutes from his watch, you would not be inclined to pay him. What he means is that he will sell you a half-hour's worth of his expertise as a lawyer: the expertise is the commodity and time is one of its measures. In a computer, time is in one sense an external measure of progress, but in another it is the raw material (the pulses) upon which the central processor operates. In a third sense, it is the product as well. The person who designs the computer and the one who programs it are both architects of time. No lawyer or executive can juggle his time in the same way or to the same degree as a designer synchronizing the control signals of his CPU. For the computer designer and the programmer, time itself is a resource to be fashioned into something useful. They feel the same ambivalence toward electronic time that the structural engineer feels toward his materials—a combination of pride at what they can do with the time available and frustration that they cannot do more.

Progress in Circles

The computer programmer is concerned about time because he wants to get a job done: he wants to make the miniature temporal world of the machine work for him. All the elaborate mathematization of computer time comes down to the desire to put time to work. The difference between time as measured in the sequencing mechanism of the machine and time as we experience it is vast. But even so, the fractions of seconds used by instructions can add up to minutes or even hours of operating time, simply because a program may call for executing millions or billions of such instructions. The operating time is often the single most important measure of work done by the computer. When a new machine is brought on the market, the first question asked is: how fast are the basic instructions to fetch data from memory, operate upon it, and return the result? A new computer installation, a collection of several processors and storage devices, measures its productivity in terms of *throughput*, that is, how many programs can be run in a fixed period of time. Conversely, the programmer grades the success of his solution to a problem by the speed of execution of his program. It is often possible to achieve the same results with two different programs, one needing hours to execute, the other seconds, and the faster program is usually better. The use of a computer is charged on the basis of time. Except for theoretical physicists, engineers and programmers probably speak and think about time more than any other specialists in our society, certainly more than philosophers or business executives concerned with employee efficiency.

Computers have a *wait state* and an executing state, and move constantly back and forth between the two. In the wait state, the central processor is performing no useful electronic work; it is waiting to receive data or to transmit it, to be given a new program task, and so on. In this state, the processor is simply allowing its timer to pulse, spending time just as an automobile consumes fuel while idling. Idleness is anathema to those who build computer systems: their goal is to design the system so that the processor is nearly always making itself useful. They arrange, for example, to have the processor work on several programs at a time, giving a split-second's exclusive attention to each, depending upon its needs. This technique is an aspect of *time-sharing*. When one program runs temporarily out of data or requires a response from a human operator at a keyboard, the processor does not wait for the potentially slow reply. After all, the ten seconds a human operator may consume in responding represents millions of additions for even a minicomputer. Instead, it goes on to the next program and periodically returns to the first one to see whether there is now more work to be done. Despite their efforts at saving time, systems programmers cannot eliminate all waste, and a computer may spend half of its day in the wait state.

Throughput, time constraints, and time-sharing are examples of the way computer specialists divide, manipulate, and sell their special commodity. These are the architects of electronic time, building temporal structures to suit their computational needs. Learning their techniques, acquiring a new and highly abstract conception of time and a driving sense of progress, is part of the apprenticeship of each new programmer.

Let us look more closely at how a programmer measures prog-

ress in his program. The central processor marks time in nanoseconds, but the programmer generally avoids these vanishingly
small fractions of seconds and speaks instead of the number of
logical instructions the machine has to execute. A program always consists of a series of instructions, which, no matter what
code the programmer may be using, is eventually transformed
into machine instructions, the basic language of the computer. I
have mentioned some typical machine instructions: add the contents of two registers and store the result, fetch a byte of data
from memory, perform a logical operation upon two bytes, and
so on. A typical computer has a repertoire of one or two hundred
such commands. Each program is a list drawn from such a repertoire, instructions to be executed in a prescribed order; the list is
always finite and seldom terribly long.

A thousand statements is a sizable program. But a thousand
statements need at most a fraction of a second to execute; in fact,
computer programs often run for minutes or hours. The reason is
that some groups of statements are executed many times over—
there are cycles built into computer programs. When the processor reaches statement 200, say, it goes back to statement 150
and executes again, performing the same operations upon new
data. Repetition gives the computer its tremendous power for
both clerical and scientific work. A program used for writing out
payroll checks will treat each employee's work time and deduc-
tions as new data, calculate the amount due, write the check, and
then repeat the same operations for the next employee. It repeats
until it runs out of data, having processed the last employee's record. A scientific program may contain several such repetitions,
for a common approach to solving a mathematical problem is
to find an approximate solution and then to refine the answer repeatedly, based upon ever-better approximations, similar to the
method high school students learned for finding square roots before the invention of pocket calculators. Such repetitions make
the computer into an electronic assembly line, processing nearly
identical units of data into output products. Without repetitions,
many if not most assignments would not be worth the trouble of
programming. For like an assembly of specialized machines, the
units of program code (slowly thought out and painfully checked
for errors) only pay off when the cost (mental or monetary) can
be "amortized" over thousands or millions of repetitions.

Computer programmers call a set of instructions to be repeated
a *loop*. The program remains in a loop perhaps millions of times

until some terminating condition is met, for example, the last unit of data is processed or the approximating answer is now considered adequate (see figure 7-1). This terminating condition is crucial. Each time the program is about to repeat, it tests to see whether the condition has been met; it loops only if there is more useful work to do.

Very often, however, the programmer makes an error and writes his instructions in such a way that the terminating condition can never be met. The result is an *infinite loop*, in which the processor repeats the same instruction indefinitely (figure 7-1). Caught in a hopeless and endless logical dance, it continues to perform the same operations upon the same unfortunate string of bits, perhaps billions of times before a human operator or a control system breaks in and calls a halt. An infinite loop ruins a program. A computer cannot perform useful work in such fashion because its mathematics, logic, and resources of time are finite.

Loops also bridge the gulf between computer time and human time. If one instruction may only require a millionth of a second, ten million repetitions of that instruction will require ten seconds, a noticeable delay for a programmer at a keyboard. An infinite loop can consume minutes of precious CPU time before the control system interrupts. Programmers try their best to avoid infinities, but loops themselves are everywhere and mistakes in constructing them frequent. There are guidelines for constructing efficient, finite loops to begin with. A programmer soon learns to conceive of his problem in just these terms: how to construct a loop with a terminating condition that will achieve a solution in a fixed amount of time.

The time limits, which the computer and its programmer confront, are of both a theoretical and a practical nature. The ultimate limit comes from the nature of the computer itself. A von Neumann machine operates in discrete steps determined by the machine's electronic timer; it can only solve those problems whose solutions require a finite number of logical steps and so a finite period of time. The number may well be astronomical, but it must be finite, and there are in fact problems that cannot be solved under this condition. This great discovery was made by Turing himself when he first created his Turing machine. Recall that such a machine consists of a tape for keeping its data and a set of unchangeable operating rules. The machine is set in motion reading the tape and stops when it has finished transforming the data into a desired output. (There is an explanation of the Turing

Figure 7-1. Flowcharts

113
Time and
Progress in
the Computer
Age

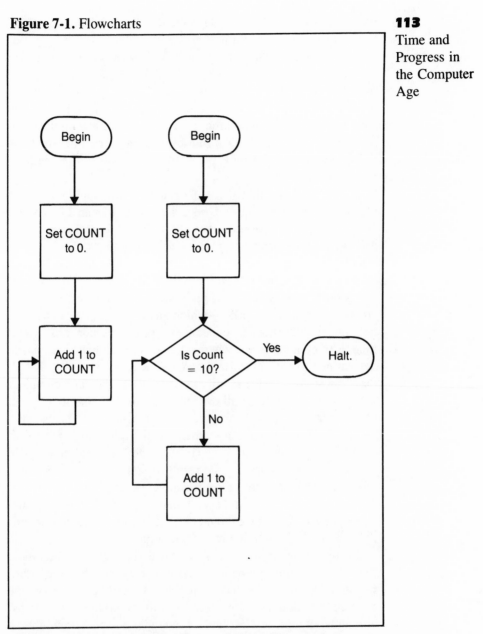

These flowcharts are spatial representations of two algorithms. Each box indicates a decision to be made or an action to be taken, and the arrows that connect boxes show the flow of control. The algorithm at the left has an infinite loop. It will continue to increase the variable COUNT indefinitely. The algorithm at the right includes a terminating condition (written in the diamond). With each repetition, the algorithm tests to see whether COUNT has reached the number 10. When 10 is reached, the algorithm halts.

machine in chapter 3.) Turing wanted to know whether any Turing machine working on any problem could on logical grounds be guaranteed to finish; he proved there was no guarantee. He showed that there are problems for which the machine may never finish its calculations, may never "halt." Such problems are said to be undecidable because they do not yield to methods of which a computer, a finite automaton, is capable. Since any digital computer obeys the same logical rules as a Turing machine, the limitations defined by Turing apply to the entire world of contemporary computers.

Do not imagine that these undecidable problems are classic philosophical issues: Does God exist? Does human life serve a larger purpose? Is the universe intelligible in logical terms? Nor are the great questions of physics undecidable: Is there an ultimately small particle or set of particles? Will the universe expand indefinitely or eventually collapse into a black hole? From a logician's point of view, all these problems are decidable because the answer to each is yes or no—which answer it happens to be is of no logical interest. The undecidable problems concern formal models of computers and whether a class of such machines can be determined to act in a certain way. What is significant is that all these questions involve infinity in some way: whether a computer working on a problem will ever finish, will ever find its way out of the loop by means of some terminating condition. Turing's theory of finite automata is a mathematical exploration of the ultimate limits of the world of electronic time. The infinite loops in Turing's programs are not the result of some programmer error, which careful scrutiny or new techniques can hope to correct. They result instead from the very nature of the digital logic machine and show that programmers must simply abandon problems they cannot hope to solve with finite logic.

Turing's work casts its philosophical shadow over the whole field of computing, perhaps more for its elegance than for its immediate relevance. The halting problem is admittedly not one the average programmer faces daily, but many common problems, although not logically insoluble, do require enormous amounts of time. Often the discrete logic of the computer does not allow for an efficient solution. For example, the computer's mathematics of approximation is sometimes hard put to deal with differential equations (involving, as they do, infinite series and limits). Programmers work very hard to overcome such limitations and to spend their computer time parsimoniously. It is not simply a question of the expense of CPU time; many problems are intractable for even the fastest contemporary machines. A program that would require a million years of CPU time (and such problems arise frequently in engineering) is not theoretically infinite but is impossible in practical terms. Wherever infinity encroaches upon the solution, the programmer must try another path. The fault may be his own if he writes his loop in such a way that it fails to terminate. The fault may lie in the problem itself. In either case, infinity threatens to shatter the careful architecture of electronic time the programmer seeks to build, to vitiate the progress of his program.

The Idea of Progress

The problem of temporal limitations takes us back again to the Greek world, whose philosophers, poets, and historians thought and wrote with their characteristic pessimism about the passage of time. That all men are mortal, limited by time, is a theme that appears as early as the *Iliad* ("Like the generations of leaves, so are the generations of men. . .") and repeatedly throughout Greek literature. The elegiac poet Mimnermus bewails the coming of old age. Herodotus, the historian of the Persian Wars, depicts King Xerxes weeping at the thought that not a single man in his splendid army will be alive in a hundred years. In contrast, seldom in Greek literature do we read of the establishment of a colony, the pursuit of a policy, the founding of an industry, or even the planting of a philosophical idea that will bring rewards in the next century. Prophesy was a common feature of life; long-term planning, at least by governments, was not. The Romans had more of a sense of history, but even they did not plan far ahead, at least in an attempt to foresee and control a changing world. Augustus's empire, once established, was meant to last forever, without more than tinkering at its boundaries and tuning of its political structure. Plato's utopias were meant to be eternal in the same static way. In the long run, time could only cause decay. The philosopher explains in detail how his perfect republic could degenerate through timocracy to aristocracy, democracy, and finally tyranny. Although his state decays in stages, there is no similar progression toward perfection, which is described in the early books of the *Republic*. The utopia could only be established by a stroke of divine intervention, enabling the philosophers to become kings or the reverse, not as a process in time.

The static character of Greek thought has often been noticed and perhaps exaggerated. Change has a role to play in many ancient philosophies, but the tendency was to associate change (motion, time) with decay and stability with perfection. Plato's beautiful definition of time as the "moving image of eternity" makes his preference plain. The real world, the world of unchanging ideas, is timeless, as it is motionless, but time belongs to the ever-changing, lesser world of our experience, a mere reflection of the ideal. We see something of the same preference for the timeless in Greek tragedy. The plays are characterized not so much by the notorious unity of time as by the abnegation of time. The passage of the day must adapt itself to the unity of the action and not the reverse, as in French tragedy, which sought to adopt the conventions of the Greeks without understanding their significance. In Greek mythology, the same situations (abduction, incest, cannibalism, and family conflict) occur repeatedly, often with the same characters. The repetition of favored themes takes precedence over the conventions of time, which must be stretched or ignored altogether. Time, when it passes at all, is often a matter of recurring situations, as in the curse that plagues the house of Atreus and claims victims in each generation.

This is significant. The Greeks were highly conscious of the fact that time is measured by cyclic motions. The tendency to think in these terms was fostered by a technology built upon simple circles: the spindle, the potter's wheel, the carpenter's lathe. But if they wanted to see evidence in nature of the cyclic character of temporal change, they had only to look at the motion of the heavenly bodies. After the work of Eudoxus of Cnidos, no ancient astronomer or philosopher seems to have doubted that there were repeated cycles in the sky. And to account for the cycles, there must be circles (or spheres in three dimensions). The universe of both Plato and Aristotle was composed of concentric spheres, and the resulting circular movement of the heavenly bodies provided the final reference for every ancient measure of time. Thus Plato wrote: "The sun, the moon, and the five other stars which are called 'planets' came into being in order to determine and preserve the numbers of time" (*Timaeus*, 38C, my translation). Aristotle claimed that the circular motion imparted to the celestial spheres by the prime mover was natural, whereas motion in a straight line required further explanation (just the reverse of the modern Newtonian view).

Circular motion was the next best thing to a state of rest. The sphere that rotates does not change shape or place; only the starry fires embedded in the celestial spheres gave any sign of its constant motion. And even then, all heavenly fires must eventually follow a circular course back to their starting places. Every celestial pattern must eventually repeat. The idea of the great cosmic year, the time needed by all bodies to return to the original positions and mark the end of one world period, was popular with various philosophers. Some before Socrates believed that the world vacillated between order and chaos. The Stoics after him taught that each world period ends in a conflagration: the slate is wiped clean and everything begins again. Plato himself sometimes hinted at such a possibility.

The ancient historians and poets tended to see human temporal events as the philosophers saw cosmic ones. Herodotus thought he could discern in the lives of individuals a pattern similar to that of kingdoms: whoever aspires to too great a share of earthly happiness must eventually suffer a downfall. This meant that all empires must decline in time and be replaced by others. Herodotus discerned that this pattern in history repeated itself as far back as he cared to and was able to look. In general, the idea of repetition throughout history competed in the ancient mind with the idea of progressive decay. The latter appealed particularly to Roman poets and orators. Cicero believed with most of his fellow citizens that the Roman republic had witnessed a steady decline in morals since the Punic Wars. The best poets all more or less subscribed to this philosophy; the image of the ages of man as a series of metals of diminishing value (gold, silver, bronze, and presently iron) was a commonplace. Virgil, among those who accepted the imperial system under Augustus, described it as a return of the age of gold:

> The great order of ages is born once again.
> Now even the Virgin returns; the rule of Saturn returns.
> Now a new lineage is sent down from high heaven.
> (Eclogue IV, 5–7, my translation)

Society here improves only by a return to a mythical condition of the remote past. There is little support among the ancients for the idea of indefinite progress in the future; that idea appears only briefly with the Sophists of the fifth century B.C. The rest of the ancient world contented itself with the reflection that conditions either remained as they were or got worse, or got worse and then returned to their original state.

Christianity offered a new, more positive view of history to the late and dying antique world, a new understanding made necessary of course by the appearance of God in history, of the advent of Christ. Unlike pagan religion, Christianity was eschatological: the whole course of human affairs had a purpose in salvation through Christ. The Christian theologian had to deny the pagan cycles of history, for other kingdoms may have risen and fallen in predictable ways, but the kingdom of God had to behave differently. Augustine wrote the *City of God* partly in response to the sack of Rome in 410, to show that the future of Christianity was not tied to that of the crumbling empire. History itself now had a goal larger than the various political orders, which must still succumb to time. Waiting for the end of time, however, turned out to be a tiresome business. Members of the early Christian community believed that it would come in their lifetime; nearly a millennium later, people were still waiting, expecting the year A.D. 1000 to be the last. Eventually, expectations came to focus more upon the historical time left to us, however long that might be.

From the Middle Ages to the nineteenth century, there gradually developed a secular view of history, a growing confidence in what humans could achieve, at least technologically. The idea of progress in history supplanted the Christian notion that man could progress only by leaving human history behind in salvation. Western Europe did not, however, return to the pagan ideas of historical cycles or inevitable deterioration. Yet the technology of the age might well have suggested cyclical theories of history. After all, the mechanisms that were helping to bring about material progress were full of cyclic gears and repetitive motions. And if the ancients saw the cycles of the heavens, the moderns understood those cycles even better, for they realized that planetary motion was even simpler and more repetitive than Ptolemy had supposed. The answer lies perhaps in the dynamic rather than the mechanical aspect of Western European technology. The windmill, the waterwheel, and particularly the steam engine were machines designed to make progress against nature by grinding grain, bailing swamps, and moving goods. The steam locomotive was the best symbol of the new view of progress, a view that developed and then declined practically in conjunction with that machine. Here was a prime mover that turned the repetitive motion of its piston and the rotary motion of its wheels into an inexorable drive forward: it represented progress in the original sense of the word. In general, the power technology of the eighteenth and nineteenth centuries suggested that mankind could literally break its way out of the cycles of history.

The idea of unlimited human progress became increasingly popular during the Enlightenment—progress through reason, which meant technology and science in the broadest sense. Many writers before had set the improvement of the human condition as the goal of science: Francis Bacon was the most conspicuous. Although he contributed little of significance to any particular science, the pages of his *Great Instauration* set out the goal in clear, if naive, terms. Those who followed in the eighteenth century (Fontenelle, Diderot, and other philosophes) elaborated Bacon's personal view into the spirit of the age. The *Encyclopédie* (1751–72) was a magnificent assertion of the achievements of technology (it included eleven volumes of illustrations of machines and tools) and of the power of education to win further gains.

There were those, like Voltaire, who sometimes doubted men were up to the task, and their voices grew more insistent throughout the period of 1750 to 1914, when the First World War ended any serious consideration of the infinite perfectibility of man or society. Rousseau's elegant diatribe against the sciences and the arts (written in 1750) set the tone for this rebellion against progress, and throughout the nineteenth century one writer after another developed his own special kind of protest. The movement in support of progress was more unified and more popular, if more shallow in its thinking, simply because it could point to the spectacular visible artifacts of technology—the steam engine, massive quantities of iron and steel, precise machine tools and instruments, improved agricultural tools, large quantities of cloth of fair quality, and so on—and to the growing prosperity of the middle class. Watches were now cheap enough to be generally available, and nearly everyone worked by the clock. Most workers did not arise with the sun, at least not in the winter, and they labored under miserable factory conditions or in mines, perhaps never seeing the sun all day.

If the whole society was regulated by the clock, and ultimately by mathematical, Newtonian time, that meant as well that the whole society enjoyed the promise of Newtonian time. Men had an infinity of time before them, an endless opportunity to develop new sources of power, expand cities, invent new machines, and extrapolate the rising curve of material progress. Although some insisted that material and spiritual progress were not the same, hardly anyone thought seriously about the fact that the material resources of the planet were limited too. The great undeveloped areas of Asia, Africa, South America, and western North America must have seemed to nineteenth-century industrialists to be inexhaustible sources of raw materials; and they might still seem inexhaustible, had it not been for the population explosion of the twentieth century.

The development of the computer could hardly be considered evidence for denying the doctrine of infinite progress. The course of Western technology itself has never given any cause to doubt it; technology has grown steadily throughout the last two centuries, and there is still no end in sight. In fact, Western technology is improving at an explosive rate. The first fully programmable electronic computer went into operation in 1949; the transistor was invented in the same year. The first laser was built in the early 1960s. All three are now common features of modern industry. In contrast, the Newcomen steam engine was first operated in about 1712, but a practical railroad system was not devised until a century later. The lesson of the twentieth century is not that technology has failed men but that men have failed to prove worthy of their technology. It was, of course, the two great wars that brought this lesson home to those who were intellectually honest enough to accept it. Men were clearly not infinitely perfectible creatures, as the Enlightenment and indeed Marxism believed them to be. But this conclusion is simply a negative one, a rejection of the old European view of progress.

Electronic technology now suggests a different view, not necessarily wiser but at least tempered by a sense of ultimate limits and characterized by a new definition of progress. For a computer's time is finite, and its progress is cyclic. Progress through repetition has in fact become a trademark of the whole industrial era, with machines of all kinds producing thousands of identical parts, and has led to the material triumphs as well as the worst excesses of industrialism. Yet the computer occupies a special place because its cycles produce not washing machines or automobiles but units of information, fragments of ideas. The computer suggests that progress through repetition makes sense not only in the industrial plant but in whole realms of logical and mathematical thought.

The computer has become the symbol of progress in our era, just as the steam engine was in the last. The steam locomotive symbolized linear progress in an age that knew no material limits, when men and women had only to push ahead with larger, faster, more powerful machines. The computer replaces the endless straight line with the loop. It aims at a predetermined goal, a terminating condition. And when it reaches that condition, it halts. Nothing good is endless in the computer world.

A program solves a problem by breaking it into small, repeatable units. Not all problems are accessible to this method, which simply means that the problems the programmer chooses, the goals he sets, are ones that *are* accessible—ones that can be handled by logical calculus, by repetition, in a limited period of time. The programmer measures progress by the ease with which his program can race through its loops in the time available. The computer mathematician seldom tries to find a programmed solution to the paradoxes of number theory, but his methods of increments lend themselves well to solving differential equations. The city planner, who wants an electronic simulation of urban pollution, divides the passage of time into units and has his program loop through the calculations, each time with updated values for industrial output and the use of automobiles. In a good program for playing chess, every statement reflects the ruthless constraints of time under which the computer will be playing. No matter how we approach the question, we always return to the fact that computer time is finite. Indeed, every resource in the electronic world (as in the real one) is limited—time, memory, central processing capacity, speed of input and output. And programming is a lesson in husbanding scarce resources.

Clearly, the realization that all resources, including time, are limited will be fundamental in shaping both social realities and the cultural and scientific outlook in the coming decades. One ecologist has recently pointed out that time may be the resource that runs out first. Economists and engineers are beginning to suggest plans for the ultimate stabilization of economic growth and for replacing expensive materials (such as metals) with abundant materials (such as glass). If we are to make progress at all, it will be progress through "recycling," a current catchword. The computer, by nature a conserver of resources, is congenial to this outlook, and it is no accident that computers figure largely in all facets of conservation and rational consumption.

Scientists too have found intriguing the idea of ultimate limits in nature. Since Einstein's general theory of relativity, they have taken seriously the possibility that the universe itself is physically limited, often compared for the layman's benefit to a four-dimensional sphere. This century has witnessed the demise of the Newtonian concepts of infinite, absolute time and space. Einstein's notion of time is mathematically beyond the reach not only of most laymen but even of intellectuals who lack a thorough education in mathematics. Still, it leads to results that everyone can appreciate: there is an ultimate speed in the universe, that of light; and, because the speed of light in a vacuum is constant, our ideas of simultaneity, mass, distance, and time itself must bend to accommodate themselves to this constant. If we add the big-bang theory of cosmology to the picture, then we have a universe with an ultimate time limit—one that began at a particular moment in the past (10 or 20 billion years ago) and that may once again collapse at some distant moment in the future.

Modern physics is altering our cosmological outlook in favor of limits, temporal as well as spatial. Electronic technology makes the same suggestions, but on a less imposing scale, and it involves more people more directly than cosmology does. After all, physicists must remain spectators in a drama whose scale dwarfs the whole earth and the humans beings on it, but programmers, even beginners, participate in the architecture of electronic time. For professional programmers, the measurement and manipulation of time becomes so ingrained in their thinking that they begin to regard human time as manipulable in the same way. Frederick Brooks has described the dangers of such thinking in *The Mythical Man-Month*. Managers of programming teams, forever trying to meet deadlines, estimate tasks in *man-months*—the product of the number of programmers by the number of months worked. This rigid scheme often leads to trouble, for it suggests that what ten men could do in three months (thirty man-months), thirty men can do in one (thirty man-months). Human effort cannot be accelerated, like the throughput of a computer system, simply by adding more "processors." Programmers collaborating on a project are not capable of such algebraic manipulation. This lesson—pessimistic for the manager, optimistic perhaps for humanity as a whole—is further evidence of a changing, less grandiose sense of human progress. We are becoming aware of our own temporal limitations to create or sustain our complex systems, technological or social.

In short, although the view of time inspired by the computer is mathematical, like that of nineteenth-century Western Europe, the view of progress differs from that peculiar mania of the nineteenth century. The computer specialist is not likely to be entirely sober about the future because he works in a field undergoing remarkable growth. We need only remember Turing's prediction that computers will successfully imitate humans by the year 2000. Some even regard the computer as a way of perfecting human beings: can we not improve men and women just as we hone programs to make them more efficient? Yet these utopian urges are more than balanced, I think, by the realities of daily work under constraints of time and computing power. In the end, the idea of progress through repetition favors stability over growth. Successful programs, like all other engineering projects, need carefully prescribed goals. Computer specialists, along with other engineers, are sympathetic to efforts at conservation and control, in society at large as well as in their own work. The experience of programming, shared by millions of educated people, is helping to change our culture's view of progress and perhaps its view of the process of history itself. Both Marxism with its industrial eschatology and capitalism with its infinite economic growth express their faith in the direction of history. May we not be returning to the ancient view, in which the cyclic rise and decline of societies seems more reasonable, and even more natural, than any promise of indefinite progress for mankind?

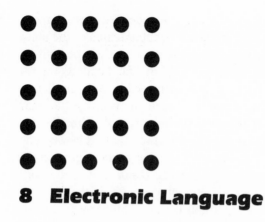

8 Electronic Language

The computer treats language as it treats logic, space, and time—with an odd combination of practicality and philosophical abstraction. On the practical level, computer languages are codes whose purpose is to represent the logical structure of problems to be solved. They are removed as far as possible from the emotional, ambiguous, and vocalized language of everyday life. And yet they are not without traditions, not without philosophical roots that reach back to the seventeenth century, and not without reminiscences that go back much further, at least to Aristotle. In creating his codes, the computer specialist in fact takes part in a debate that has been lively for thousands of years. Should language evoke or simply denote? Does it enter into a magical or at least mysterious relationship with the thoughts and experiences it describes, or is it primarily a tool for laying bare the structure of the world? Is language essentially poetic or logical? As we look first at computer language itself, then at the traditions of which it is a part, we shall not be surprised to see which side the computer favors in this debate. Poets, philosophers, and computer programmers may agree that language provides a path to knowledge. They do not agree on the nature of language or the kind of knowledge to which it leads.

Natural and Artificial Language

To speak of computer "language" is to speak metaphorically, although it is a metaphor dying from overuse. Even those who

never come near a computer have heard of FORTRAN, COBOL,
or PASCAL and know that each is a cryptic language in which
one speaks to the machine. FORTRAN (the name stands for
"Formula Translator"), PASCAL, and the others are not living
languages, of course. Programmers themselves realize that when
they distinguish between artificial or programming languages and
"natural" ones, such as English. The designation "artificial" itself says something about the programmer's outlook. English is
also the product of artifice, but its creation through centuries was
not self-conscious and rational, as the creation of programming
languages is.

Programming languages are not spoken, and this too is an important difference. They are not designed to be understood in oral
communication, for current computers are poor at making sense
of sound waves. Even if it were possible to speak to a computer
(which is at least conceivable), few programmers would want to
conduct a dialogue in FORTRAN. Here the image provided by
science fiction is accurate. When humans speak to their robots
and electronic brains, they do so in something approximating English, often omitting articles and other small words to suggest the
computer's preference for reducing language to the bare bones of
logic. Speaking FORTRAN would be as difficult as speaking algebra because the relationship that such codes establish among
their symbols can be grasped only when laid out in space and examined with the eye. The eye can peruse a line several times to
tease out its meaning or jump back ten lines to check the definition or use of a symbol. For most of us, a spoken version of a
FORTRAN statement disappears as soon as it is uttered. Even
two skilled programmers must sit down with a written version if
they wish to discuss a program. The same holds for mathematicians and their equations; it is truly astonishing when we read of a
blind mathematician such as Euler, who could somehow visualize
his proofs without the aid of paper. Natural language works differently: it evolved to be fully intelligible in speech because spoken language preceded writing by millennia. It is highly redun-
dant, unlike good computer programming. In English we phrase
and rephrase ideas to get them across; we do not assign one univocal symbol to each idea and expect our audience to remember
it for the rest of our conversation.

FORTRAN has much more restricted uses than English. It is
not capable of expressing emotions, or indeed many reasoned insights, but only a narrow range of logically defined problems.

Nor does a programming language evolve in the same unrestricted way that a natural language does. English is in constant flux. As some words fall into disuse, others come into fashion; grammar changes, and pronunciation varies from one region to another, from one group of speakers to another, and over time. Most of these changes are unplanned and even dismaying to the literary establishment. Change, especially in pronunciation, is for most speakers wholly unconscious, and it is therefore democratic in the best and the worst sense—a chaotic process in which collective change is the sum of millions of idiosyncrasies. Language reform by committee is usually doomed from the outset; however, modern Hebrew may be an exception.

Computer languages, on the other hand, change in just such an autocratic fashion—by decree from the management. A programmer is not free to modify FORTRAN as he wishes because any deviation in the rigid syntax expected will simply cause his program to fail. Change is never spontaneous. A group of programmers, dissatisfied with some small aspect of the language, may agitate for reform at their computer center. The reform comes before the management of the center, and, once a decision is made, some systems programmer is given the job of rewriting the program that controls FORTRAN. Greater reforms may arise in adapting a language to a new line of machines. According to the institutional anarchy of committees, each computer language develops dialects as it spreads out through North America and Europe. Soon enough, many competing versions exist, not entirely compatible with one another but all recognizably the same language.

Computer languages, like natural ones, can be grouped into families. The Indo-European grandfather of electronic language is FORTRAN: PL/1 and the business language COBOL have descended from it, for both of these refine its syntax to meet other needs. Other more exotic language groups begin with other syntactic principles. All the general-purpose languages have approximately the same power: what can be programmed in one can be done in the others, because all of them are at bottom instructions for building a Turing machine. Each language, however, has its own color, reflecting the philosophy under which it was designed. FORTRAN (now well over twenty years old, a remarkable longevity for computer software) remains the closest thing to a lingua franca: it is straightforward, not particularly subtle, an engineering language. ALGOL is an elegant European language that

PASCAL is its younger cousin. PL/1 is a sprawling language that attempts to provide all the facilities that any programmer (scientist, engineer, or businessman) could want; seeking to please everyone, it pleases no one. LISP and APL are concise and beautifully logical languages, popular with those (such as artificial-intelligence specialists) who want to emphasize the derivation of computer programming from symbolic logic.

Men of letters used to speak in similar terms of natural languages as vehicles of expression—of the heavy, philosophical quality of German, the delicacy of French, the clarity of Latin, and so on. A sense of style does indeed enter into both the choice of a programming language and the writing of the program itself. Computer programmers, like mathematicians, stress elegance in their work. There are programs made simply to execute and others that have been polished to be not only more efficient but also clearer to read and easier to modify.

A program is written for two very different kinds of readers: the computer that executes it and other human beings who may need to read and revise it. For this second group, the programmer inserts comments in English, spaces the instructions, and tries to keep the program as straightforward as possible. Designers are forever trying to make programming languages more natural and more accessible to laymen, for economic as well as technical reasons; the business world would be happier to bypass programmers and put its own clerks and executives in close contact with its computers. The fact remains that today's programming languages are codes rather than natural languages: human workers must meet the machine far more than halfway. The computer retains much of its mystery precisely because its medium of communication is a code, hard to decipher and notoriously hard to remember. Computer programs written by one specialist are usually unreadable by another without copious comments in English, and every programmer has had the frustrating experience of being unable to decipher code that he himself has written a few weeks or months before.

The Hierarchy of Computer Language

To bridge the gap between its two readers, machine and programmer, computer language is designed hierarchically. Recall that the central processing unit of a computer responds to a set of several hundred instructions, known appropriately as the *machine language*. This is fundamentally the only command code that the computer understands, a code more or less built into its hardware. Consisting of 1s and 0s like everything else in the computer, machine language is extremely cumbersome. If a programmer wants to write an instruction that adds the contents of one word of memory to a second, he must look up, for example, the eight-digit symbol for this add instruction in a table, as well as the addresses of the two words. The result will be a string of perhaps thirty-two digits, which he can easily mistake in copying. A program of several dozen such instructions would be a nightmare to write, to proofread, and to correct. Nevertheless, engineers did work directly in machine language in the first years of computing. Then they began to develop more readable codes, taking advantage of simple mnemonic devices. These new codes were called *assembly languages* and are still in use.

The principle behind assembly language is that a human programmer can remember names more easily than numbers. Short, fixed names (such as ADD, SUB, MUL) are given to all the machine instructions, and programmers are allowed to compose their own names for storage locations, that is, for variables in the mathematical sense (names like A1, SPEED, SUM). The computer still executes instructions only in machine language, but with the help of a previously written program, it now translates automatically from assembly into machine language and then executes. That is, it transforms the program written with names into a long string of binary digits that the processor understands. This translation program (itself called an *assembler*) saves the programmer the trouble of looking up binary codes in a table and writing them out himself. In general, the programmer still writes one statement in assembly language for each instruction he intends the machine to execute. He is still very much tied to the logical structure of the machine he is using, although he is now writing at one remove from the machine language.

The next step was to develop codes that removed the programmer further from his machine, allowing him to write in a more mathematical language. These codes are called *high-level languages* and are translated into machine instructions not by relatively simple assembler programs but by complex programs called *compilers*. The first such language to be used widely was FORTRAN in the late 1950s. There have been dozens since. In

FORTRAN, the programmer writes instructions that look much like algebra: for example, C = A + B. In other languages, the statements may look more like symbolic logic or even simple English. Each FORTRAN statement calls on the CPU to perform a number of elementary operations, so each must be converted into a number of instructions in machine language. A compiler program has precisely this task, accepting FORTRAN statements as its input, analyzing them into their constituent parts, and generating machine statements as output. Like a human translator, a compiler listens in one language and speaks in another. Unlike a human, who brings to the task of translation knowledge about the meaning of words and the likely intentions of the speaker or writer, the compiler knows nothing of the larger purpose of the FORTRAN program or the intentions of its programmer. No one would claim that he understands French if all he could do was to identify the subject and predicate in a French sentence, but a compiler understands FORTRAN in just this sense—it can analyze the syntax of statements in FORTRAN. For this reason, expressions written in FORTRAN, or in any other high-level language, must be structurally unambiguous. The compiler cannot make choices among alternative analyses; it lacks by definition the human translator's ability to interpret.

Here, then, is a cardinal quality of computer language: its layered or hierarchical structure. Computer codes are classified in terms of their distance from the binary language of machine instructions and their proximity to the traditional languages of mathematics and logic. On the highest level are such compiler languages as FORTRAN; below them are the assembly languages; below these are the machine instructions (figure 8-1). Actually, even higher levels are possible. A program written in PASCAL, for example, may accept more natural English expressions, such as "multiply distance by velocity," and convert these into machine operations. In this case, the PASCAL program is itself a compiler for the English language that it is equipped to process. A compiler mediates between a high level and a low one. The terms "high" and "low" seem prejudicial: the high-level language is somewhat closer to the English-speaking user, although our own language towers in complexity and richness over anything the computer can yet process. On the other hand, computer language is made meaningful only by its execution. Executing a FORTRAN command allows it to realize its meaning in action. As with any hierarchy, the units at the top give the com-

Figure 8-1. Hierarchy of Computer Languages

High-level Language (FORTRAN)	Assembly Language (hypothetical)	Machine Language (hypothetical 16-bit)
VEL = 10	LDI 10	0000110000001010
	STA VEL	0001000100010000
POS = VEL * (TIME + INC)	LDA TIME	0000100100010010
	ADA INC	0010000100010100
	MUL VEL	0011100100010000
	STA POS	0001000100010110

Three levels of electronic language are shown here. One command in a high-level language might translate into two or three (or many!) in assembly language. Assembly language still permits the use of names (VEL, POS, TIME, INC) and the use of mnemonics for instructions (LDI, LDA, ADA, MUL, and STA are all opcodes, or names of machine operations). In machine language, even these must be replaced with strings of binary numbers.

mands, intermediates pass commands along, and units at the bottom carry them out. The lowly machine instructions are the only ones that actually perform computations.

Compilers and assemblers are translation programs: they accept as input coded statements at one level of the electronic hierarchy and produce output instructions at a lower level. The process of translation may be terribly complex, but it is not mysterious. It is accomplished by an algorithm; no intuitions are required. For the compiler, to "understand" a statement in FORTRAN is simply to process that statement step by step into an executable form. After its execution, a statement exercises no further influence on the rest of the program. In English the mean-

ing of a sentence can change radically because of the sentence
that follows it, for the sentence remains active and resonant in memory long after it has been read or uttered. Each statement written in a computer language, however, commands the complete attention of the machine for the brief moment of its actual execution; then it ceases to have meaning unless (in the case of a looping program) it comes around again for execution.

Furthermore, because computer language has meaning only in action, no ambiguity can be tolerated. If a command in FORTRAN has two possible interpretations, the compiler would have to generate two sets of machine instructions. A central processor, however, can execute only one instruction at a time; and it cannot choose freely between the two sets of instructions. Thus computer language is at every level univocal: each statement is either entirely clear or simply wrong, and to guarantee its clarity, the language possesses a rigid syntax of permissible expressions. This very rigidity means that programmers, who do not think so consistently by nature, often make minor errors, leaving out punctuation or parentheses, or misspelling. When the compiler comes upon a statement that does not conform, it may try to guess at what the programmer meant to write—did he omit a comma or semicolon? But the guessing is limited because the compiler never wants to choose between operationally different meanings. In the end, the compiler must simply pass the offending statement by and omit altogether the machine instructions it might have generated. Usually the program cannot be executed at all. In short, the ambiguity that is so important to human communication is fatal to the computer.

Ambiguity in English is in many cases the difference between language and intention, for a multiplicity of meanings may be present in a single English sentence. In logical terms, our spoken and written language often fails us, not making our intentions clear to others. On the other hand, this failure is one of the qualities of language that makes poetry possible and in general allows us great economies in communication. There are times when being ambiguous conveys exactly the right sense. The message, however, often concerns emotions or intuitions, which computer languages are not designed to represent. Electronic representation as a series of symbols on punched card, magnetic tape, or magnetic disk is a flawless representation (within the error tolerance of the particular system). But a FORTRAN statement is nothing other than such symbols. In a natural language, the written expression is only a part of the whole language, and not always the most important part. Any English sentence on paper can be spoken in half a dozen ways, producing as many nuances of meaning. Every English noun has a set of connotations that color the context in which it occurs. There are no connotations, no fuzzy meanings, in a FORTRAN variable. Uncertainty in a computer language will not produce poetry, express emotion, add color, or do anything at which natural language excels. It will simply produce an error, and the program will have to be repaired.

Electronic languages are designed by mathematicians and logicians as an instrument for solving technical problems. Such people may not always appreciate the ambiguities and nuances of literary English. If they do, they cannot incorporate these nuances into their compilers, for the logical nature of circuits and storage registers makes ambiguity impossible. The only languages appropriate to an electronic system are ones as precise and structurally simple as FORTRAN and LISP. The only definition that makes good sense is the operational definition: translate the statement into machine language, execute, and examine the result. Finally, there is no such thing as a thought in FORTRAN that cannot be expressed in a FORTRAN statement; in the domain of the computer, thought and language coincide.

Poetry and Logic

In the computer world, thought "descends" to the level of language. Computer language consists of strings of arbitrary symbols; computer thought is nothing more than the manipulation of these strings according to the rules of logic. There is of course an alternative view, richly represented in ancient, medieval, and modern literature, that language is more than the sum of its syntactic parts and may serve as a pathway to higher realms of thought or being. Giambattista Vico, that erratic genius of eighteenth-century political philosophy, said that the first language among the gentiles was poetry and the first wisdom poetic wisdom. Although no one today believes that Greek contemporaries of Homer spoke in hexameters, it is true that the early Greek attitude toward language could fairly be called poetic, even religious or magical. With the fifth-century enlightenment in Athens, a new view challenged the old as the so-called Sophists began to use words in a more coldly logical fashion, as counters to be manipulated for solving rhetorical or philosophical problems. From that time on, these alternative uses of language—the logical and the poetic—have competed for primacy in each age. For some writers (I think immediately of Plato), the logical and the poetic existed side by side in a state of uneasy truce and occasional hostility. For most, however, one or the other view of language prevailed and determined their intellectual horizons.

Both views begin with the premise that words are symbols that stand for something beyond themselves, beyond sounds in the air or marks on paper (clay, stone, or magnetic tape). For the poetic mind, the symbol stands in an immediate and natural relation to the thing symbolized. God may have ordained the names for things or man may have chosen them, but in either case the names fit. Words have a power over things. They do more than denote objects; they control them. They not only name ideas like truth or goodness but also lead us to the ideas they name. Now the logical mind will have none of this. The act of symbolism is one of pure invention, and words are related to things only by convention. They allow us no magic control over the world of objects; it is only because we live in a certain culture with a certain language that we use the names we do. On the other hand, the way words fit together, the structures that we create when we use language, is itself of great interest to the logical mind, apart from any knowledge of the world outside gained through language.

In general, the poetic mind prefers the oral language to the written. Primitive cultures, which lack the art of writing, are most likely to accord great power to the spoken word, particularly to names: hence their taboos about the names of God, the need to keep one's name secret from enemies, and so on. In almost every culture, primitive or advanced, spoken words have a greater sensual and aesthetic impact than written ones, and poets have always insisted that their work will yield its full effect only when read aloud, recited, or, in the case of drama, played to an audience. Poets even in our day are men and women who half believe in the magic power of language to assert itself in the world of things. Sound, after all, does reach out to us, does force itself upon the world, in a way that written sentences do not.

As McLuhan and others have shown us, writing or printing fosters a more logical attitude toward language. When the reader is no longer bombarded by the words, he can distance himself from his text. He has time to reflect, to reread, to analyze. Writing and printing are themselves analytical processes that break down the stream of spoken language into discrete units—alphabetical symbols, words, sentences. The reader uses his eyes as well as or instead of his ears and is in every way encouraged to take a more abstract view of the language he sees. The written or printed sentence lends itself to structural analysis as the spoken does not because the reader's eye can play back and forth over the words, giving him time to divide the sentence into visually appreciated parts and to reflect on the grammatical function. Silent, critical reading also impresses upon us the arbitrary quality of each symbol. We have all had the experience of suddenly seeing in a new light a quite ordinary word on a printed page. We have read the word thousands of times since childhood, yet suddenly that particular arrangement of letters seems utterly arbitrary, and we realize that these letters could stand for anything we care to define. Such an experience is a precondition for the logical view of language.

The Ancient View

This distinction can be applied to the culture of Greece and Rome. Vico was right in claiming that the wisdom of the ancients was wisdom embodied in poetry before logical prose. Ancient civilization had in fact created a number of poetic masterpieces before its philosophers began to write logically. Archaic Greece from Homer to Aeschylus used the new art of recording and fixing words (their alphabet borrowed from the Phoenicians) only sparingly. It was still an oral society. Whether or not he himself was literate, Homer certainly composed his epic poems to be recited before an audience, possibly a grand audience on festive occasions. The lyric poets of the archaic period also expected to have their poems read aloud, if not sung. Even such philosophers as Pythagoras and Heraclitus relied on immediate contact and oral teaching, and for this reason early Greek philosophy was full of imagery and epigrams that we now associate with poetry. Greek tragedy, like all drama, was not meant to be read but to be played before an audience.

The age was alive to the incantatory and resonant qualities of the spoken word. The Greeks had already progressed far enough to be free of the grosser superstitions of primitive cultures but not so far as to forget the primitive's admiration of the power of language. In the Greek epics, characters speak "winged words"— suggesting that words themselves and the ideas they embody are as real as birds and spears that also fly through the air. Two hundred years later, the audience of Aeschylus's plays, if not Euripides', could still manage to believe that a curse uttered upon a king could bring destruction to him or his progeny.

By the fifth century, however, the technology of writing, together with other cultural forces, brought forth a new point of view. The infamous Sophists, of whom Plato was so critical, were itinerant orators and philosophers who specialized in verbal pyrotechnics and cynical attacks on the established order. These Sophists relied as much as anyone on the power of the spoken word, for they were debaters and orators. Most of them, however, taught their students to regard language as something to be manipulated arbitrarily for their own purposes, whether philosophical or monetary. For the Sophists, words had lost some of the awesome power they possessed for earlier generations of thinkers.

The sophistic attitude was too much for Plato, who in some ways reverted to a more poetic view of language. For Plato, a supreme prose artist, spoken language remained the key to philosophy because it was the technique of Socratic dialogue (spoken question and immediate reply) that opened the mind to philosophical issues. As Harold Innis rightly says, the attempt to translate Socrates' oral teaching into a literary form made Plato aware of the destructive impact of writing upon the old oral culture (*Empire and Communications*, 56). In the *Phaedrus*, Socrates reports a conversation between the Egyptian god Theuth, the inventor of writing, and Thamos, the wiser if less inventive deity. Thamos points out that Theuth's invention "will encourage forgetfulness in the minds of your pupils, because they will not exercise their memories, relying on the external, written symbols rather than the process of reminiscence within themselves" (*Phaedrus*, 275A, my translation). Writing, Thamos continues, is like painting in that it provides the semblance of wisdom rather than the reality.

Plato could write with such mistrust about the technological revolution because he understood what had been lost as well as gained. The external symbol, the abstraction, replaced the resonant act of memory that came from within. Plato never regarded any facet of the world as arbitrary, and he perhaps realized that the abstract, distancing quality of the written word could lead to a

Electronic
Language

theory of language as arbitrary symbol, a theory that Aristotle and the Stoics later developed. Spoken language had a philosophic function, as did silence itself: question and answer brought a philosopher to a certain height, from which he could only proceed to contemplate the forms of truth and beauty in silence. Language was not thought for Plato; rather, it led to great philosophical thoughts.

When Aristotle took the step of denying Plato's eternal ideas, he removed the anagogical function of language. The way was clear to treat language in strictly logical terms, to emphasize the conventional quality of names and to examine linguistic structure. Aristotle divided words into categories that were at least partly based on their grammatical function. He maintained that analogy, and therefore convention, was the guiding principle of grammar and etymology. He also classified syllogisms, providing the most explicit ancient statement of the relation between language and logical thought. The syllogistic "figures," which classify the types of premises that lead to valid conclusions, were a way of looking beyond the meaning of sentences to their bare logical form. Aristotle went so far as to use letters to represent names and properties in the abstract. One such mood (the one called "Barbara" by the medieval logicians) runs thus: if A is predicated of all B and B of all C, then A is predicated of all C. This form may be fitted like a template over an endless number of sentences in natural language, allowing the logician to see at a glance what these sentences have in common. After Aristotle, the Stoics revised his logic, making it even more abstract and rigorous. They supported the view that language is created by analogy and spelled out in the clearest possible terms the distinction between the word (the linguistic sign) and the object denoted by the word.

Still, the ancients never achieved a fully symbolic logic, nor did they ever come to see language as a fully arbitrary syntactic structure. Indeed, their technology for preserving language worked forcefully against such trends. Even after the fifth century in Greece, the rate of literacy remained low, and books never became the ubiquitous artifacts of culture and even business that they are today. Ancient writings had to be copied laboriously by hand, and, at least until well into the time of the Roman empire, they were not the convenient books we know today (in which it is easy to flip exactly to the desired page) but rather rolls of papyrus, in which the reader had to unravel his way to the passage

he needed. In the anachronistic terms of electronic data process- **137**
ing, the papyrus roll was a slow, linear-access device, like a mag-   Electronic
netic tape, but the book allows random access and so makes in-   Language
formation more accessible.

An ancient scholar or philosopher with a large library would
choose to memorize as much as possible to avoid the cumbersome shuffling of rolls. In addition, words on an ancient page
were written without word division and with little or no punctuation. To understand such a mass of letters, the ancients had to
read aloud. Just as most of us today must hear a musical score in
order to make sense of it, so the ancients could only make sense
of a text by using their ears. Because the sound of language was
never eliminated from ancient reading and writing, Greece and
Rome remained largely oral cultures, in which books served as
means of preserving for future generations the voices of the past.
Indeed, "Greek civilization was a reflection of the power of the
spoken word" (Innis, *Empire and Communications*, 56).

In this context, it is striking that Aristotle and the Stoics went
as far as they did in the logical analysis of language. Grammatical
studies were best pursued in the silent, visual world of the post-
Renaissance university, not in the noisy ancient library, with everyone muttering the words of his text, or in an ancient banquet,
with a slave reciting the text to the guests. Even the avowed anal-
ogists among ancient philosophers had difficulty freeing themselves from the grip of oral culture, in which words act as guides
into the world of nature and the world of ideas. If they had managed to free themselves, they would have invented their own
symbolic language to make logical and mathematical manipulations easier, as the moderns have done. Yet even Aristotle and the
Stoics could not get beyond using single letters to stand for names
and properties; they never thought to invent symbols for such
logical operators as *if . . . then* and *or*. Nor did Greek mathematicians realize the advantages of such symbolism; they relied
instead in their proofs upon the tedious repetitions of words in
ordinary language.

The Greek technology of language was a manual technology.
As with other ancient crafts, there was no automation, no distancing of the craftsman from his work through machines. The
scribe was in immediate physical contact with the book he produced; he felt the letters form under his pen. His tactile and muscular participation in making the words was far greater than that
of a modern typist. The scribe also repeated the words as he cop-

ied and so brought a third sense into play in the craft of writing, as McLuhan has emphasized. Words that were felt and heard as well as seen had an immediacy and reality that we can scarcely appreciate today. Abstraction in reading was similarly discouraged by the run-on style of writing and the need to vocalize each word.

Altogether, the ancient scribal culture was inclined to regard a written page as a palpable texture, a pattern of words that reproduced the patterns of the larger world beyond the page. Hence Plato's comparison of writing to the art of drawing. Elsewhere Plato even used the metaphor of weaving (another manual art) to explain how a name reproduces reality. When Socrates suggests the definition "A name, then, is a sort of didactic instrument that separates out reality as a shuttle separates fabric on a loom?" his interlocutor agrees without hesitation (*Cratylus*, 388B-C, my translation).

The Western European View

Medieval culture perhaps remained closer to the ancient world in its attitude toward language than in its attitude toward space, time, or history. In philosophy, the ancient dispute about the relationship of things to names was translated into the well-known realist-nominalist controversy, still with reliance on ancient authorities. In general, the culture still regarded language as a collection of names, and the resonance and supernatural power of names impressed the medieval mind just as they had the ancient. The mass, a litany of such words as bread and body, wine and blood, was a daily reminder of the magic of spoken language, magic as real to a medieval theologian as it had ever been to Aeschylus. After all, to the theologian, the appropriate word could bring men into physical communion with God; another word could mean eternal salvation.

One reason that the Middle Ages did not emancipate itself from ancient notions of language is that its technology of language was fundamentally the same as the ancient. With its mechanical technology—waterwheels, windmills, and clocks—the Middle Ages had surpassed or at least diverged from the ancient world, but books still had to be copied by hand. The medieval scribes had adopted the innovation of later antiquity, writing exclusively in codices (paged books) rather than lengthy rolls. The essence of the craft had not changed. Every manuscript was a

fairly costly creation, full of its own particular errors, variant readings, and scholarly notes made by its owner or owners. Every manuscript was unique. And, because of the variation among handwritten letters, reading itself was as slow and difficult an art as before. The medieval scholar, like his ancient counterpart, read aloud, bringing the words of his manuscript to life with his voice. Books were rare and precious in the Middle Ages, as indicated by the fact that a university lecture was usually what the Latin origin of the word suggests—a complete reading aloud of the text. The professor would often have the only copy, which he would dutifully read out to his students, while adding his own running commentary.

A change in thinking about language came later, in the Renaissance, and an obvious force for that change was technological, the invention of printing. Writers as early as Francis Bacon understood the impact of technology upon language. The printing press did not simply change the mode of book production; it changed the literary community's perception of language and the acquisition of knowledge. Books were no longer so scarce, nor was any book unique. The printing press turned out thousands of more or less exact duplicates of a typeset page. There was more likelihood of catching an error on a printed page than on a hand-copied one, and what errors remained belonged to every copy. Borrowing a friend's manuscript to collate errors and compare marginal notes was no longer necessary. And printing was the first industry to organize for mass production. As early as the fifteenth century, the careful handcrafting of a book became obsolete (this applies only to the printing of the text, not to binding, which remained a charming craft for much longer). Hundreds or thousands of identical literary products rolled off the assembly lines in Venice, Nuremberg, Mainz, Basel—centuries before Ford dreamed of doing the same with automobiles. McLuhan sums it up well by saying that "the invention of typography confirmed and extended the visual stress of applied knowledge, providing the first uniformly repeatable commodity, the first assembly-line, and the first mass-production" (*The Gutenberg Galaxy*, 124).

The mechanization of bookmaking changed the very art of reading. At first printers imitated manuscripts in their layout: the choice of type, the ligatures, and the abbreviations made the first printed books as aesthetically pleasing and as difficult to read as manuscripts. But soon enough, standard letterforms evolved, and reading a streamlined printed text was quicker and less strenuous than reading the most tidy medieval codex. Indeed, quicker reading became a necessity because printing vastly increased the number of books. Where medieval and ancient men had read aloud, working their way vocally through each word in the text, the new post-Renaissance reader worked silently, bypassing his ear altogether and telegraphing the message to his brain. This is the method we are still taught today; silent, less evocative, more efficient, it allows us to "process" six hundred words per minute, whereas the medieval scholar could scarcely achieve two hundred.

Vision became the primary sense for taking in knowledge communicated through language, and for a variety of reasons this change encouraged a more abstract, more theoretical view of language. The spoken word has an immediate effect upon us; indeed, unless it remains to resonate in our memory, it dies as soon as the vibrating column of air passes our ears. But for an instant, the spoken word lives in a way that the printed word does not. Poets of any age remind us that their words are meant to be heard as well as seen. Even today, they adopt an ancient or medieval attitude toward names and language and demand from their audience some concession to that same attitude. It is a concession we make relatively seldom. Since the invention of printing, thinking men and women have spent more of their time reading silently, reading for content, and less of their time listening. The silent reader is more inclined to regard words as lifeless symbols whose job it is to communicate a message. Spoken words are of course symbolic too—patterns of sound waves that speakers of one language agree to have certain meanings. This way of looking at aural communication is quite recent, however. Only in the last few centuries have physicists understood sound waves well enough to regard them as a medium of communication, and the careful analysis of phonemes belongs only to twentieth-century linguistics. Only when the printed word freed itself completely from sound did it become natural to regard words as arbitrary signs of the ideas they called to mind. In the centuries following the invention of the printing press, interest in the power of symbols of all kinds grew remarkably.

The abstraction of the art of reading also made men more aware of the structure of language. The grammar of a spoken sentence is hard to analyze and communicate to others, for the words run away as they are uttered and leave behind only memories. A written sentence is permanent; the reader can view its several parts at the same time or he can refer from the predicate back to the subject. The silent reader is more likely to take in groups of words at a glance and to see noun and verb phrases rather than isolated words. Not surprisingly, silent readers in the Renaissance and after found structures in language that were seldom noticed before. If the idea of a universal grammar valid for all languages began in the Middle Ages, it flowered in the seventeenth century with the Jansenists of Port Royal. And long before that, humanists in the tradition of Ramus and Erasmus were applying a new awareness of grammatical construction to their study of the ancient authors.

I am not suggesting that the new feeling in the West for abstract symbols and the new emphasis on the structure of language were due solely to the mechanization of the book. In fact, printing only prepared the way for another event of equal importance: the mathematical revolution of the seventeenth century. Since the time of the Greeks, mathematics had been the paradigm of pure, abstract thought. Mathematicians had always used language in a way precisely opposite to the evocative language of the poet. Plato admired mathematics for just this reason: it pulled the mind away from concrete, earthly concerns to the higher realm of ideas. Ancient and Western European men had always credited mathematics with abstract beauty.

Physicists of the seventeenth century, however, made their contemporaries grant another more stunning quality to the abstract science when they demonstrated, as Galileo put it, that the book of nature was written in the language of mathematics. The beauty of mathematics had previously been thought otherworldly. When Aristotle built his description of the physical world, he deliberately avoided anything more complicated than simple arithmetic and intuitive geometry, so the prevailing physics of the ancient and medieval world had been logical and philosophical but not mathematical. Aristotle was reacting against the excesses of Plato and the Pythagoreans, who had turned cosmology into number theory and tried to make the world fit their mystical notions of geometry rather than fitting their notions to the world. Galileo and his followers indeed looked back to the Pythagoreans and Plato as spiritual forerunners, but their far more pragmatic approach led to a success that the Greek mathematical philosophers had not dreamed possible. The seventeenth century saw the development of analytic geometry and calculus, new mathematical tools for describing the movement of matter in space; these went far beyond the static bias of Euclidean geometry.

This new mathematics proposed to analyze nature at a deeper level than ancient geometry had done. To do so, Galileo and his followers found it necessary to abstract and simplify, to drain experience of all color, smell, taste, and other "secondary qualities" in order to reach a logical core that was amenable to their equations. The irony was, then, that mathematics had made itself more efficacious in this world precisely by becoming more abstract than ever. Before the seventeenth century, mathematics was not a language of its own, but at most a dialect, a scholarly jargon. It had its vocabulary, as did the other liberal arts, but a mathematician wrote his proofs out in words that any other scholar could read, if not understand. Galileo is still largely a verbal mathematician, but his is the century that finally realized the power of symbolism. Descartes and Leibniz led the way in making algebra and calculus a matter of x, y, and symbols of a higher order such as d/dx and d/dy. Their mathematics did become a language, a truly artificial language that was available only to a privileged few.

The new language had a range of expression limited to the bare world of primary qualities, such as mass and distance. It was in fact the first and by far the most successful attempt at "Newspeak," for by definition one could describe in the new symbols only those aspects of experience that yielded themselves to mathematical equations. With verbal mathematics, there was at least the danger that occult forces or such qualities as color and texture could work their way into the argument. Freed from these distractions, the mathematical physicist explained mechanical forces in nature as never before. Galileo's telescope, by the way, was a perfect symbol of this new spirit of the mathematical physicist, who purposely narrowed his field of vision in order to see more sharply and further than before.

Whatever the limitations of this new vision, it could not be denied that the language of mathematics was spectacularly successful in those areas of experience that yielded to quantifying and symbolic manipulation. Then as now, the issue was to establish what areas did yield. Some thinkers were captivated by the qualities of mathematics and logic as vehicles for ideas and sought to reduce much or all of human experience to a purely logical calculus, based on the traditions of formal logic and inspired by the

great success of the new mathematics. The seventeenth century
saw the flowering of the movement to create a "universal character," a language that men could use to communicate more easily
and indeed to think more clearly. Many in the movement had the
practical goal of replacing the scholar's tongue, Latin, with another more rational lingua franca. Around this time, increased
contact with China gave some the mistaken impression that the
Chinese possessed a system of writing that expressed ideas pure
and simple, that each of their beautiful symbols represented one
fundamental notion. Whole dictionaries and grammars were
composed for new ideograph systems that were intended to be
rational and easy to learn.

Leibniz had a grander plan than merely an ersatz Latin: he
wanted to create a logical language with the certainty of mathematics, one that could extend its discourse over the whole range
of human problems, in particular, metaphysics, theology, and
ethics. The idea was to begin in the Cartesian spirit by inventing
symbols for a relatively few primitive notions and a set of rules
for combining these symbols. Sentences in such a language
would each follow from the last by the rules of logic, and the result would be discourse like that of twentieth-century symbolic
logic. However, with such a universal logical language, proofs
would go far beyond those our logicians now produce. A philosopher could prove once and for all the existence and nature of
God, the world, and virtue; indeed, all religious and philosophical disputes would be settled, according to Leibniz, with the proposal "let us calculate."

As compelling as the idea was, even Leibniz with his remarkable gifts could not get beyond suggesting such a project. All
such universal characters have failed to the present day, although
computer specialists in artificial intelligence are still trying. Nevertheless, the popularity of the universal character in its day (and
the fact that such thinkers as Descartes and Leibniz were charmed
by it) indicates the exuberance, the excitement, generated by the
new mathematical view of language. In itself, Leibniz's plan to
calculate a proof for the existence of God could have very little
influence. But the assumption behind the plan had a great future—the assumption that all thought was ultimately reducible to
language. After all, the new symbolic language of mathematics
allowed men to contemplate the physical universe with a precision never before possible, and Leibniz hoped to bring the same
clarity to philosophical issues, to teach philosophers a language

in which they could think clearly about issues that had always been a muddle.

This is not to say that Leibniz's view of language has not been disputed. The trend toward an increasingly logical view of language is one against which many have rebelled, sometimes magnificently but without lasting success. But even linguists with a feeling for poetry and the humanistic uses of language, beginning with the Germans Johann Gottfried von Herder and Wilhelm von Humboldt, were receptive to the notion that language could foster right-thinking or at least make it possible (a mild form of Leibniz's assumption). Herder himself claimed that any higher reasoning and abstraction demanded the symbolism, the faculties of denotation and connotation, that language provides. He also distinguished artificial from natural language and was emphatic in claiming that human language is artificial, created by man and enabling him to reason, in contrast to animals, who communicate in a language given them by nature. In the twentieth century, the American linguist Benjamin Whorf championed the notion that language and thought are one. Although many linguists speak disparagingly of Whorfism today—Whorf was, they feel, no scientist of language—very few would be willing to deny the intimate connection of language and thought.

Meanwhile, twentieth-century philosophy embraced the radical position that language could and should be a fully logical calculus. In fact, philosophy in the first half of our century might well be characterized as an extended love affair with language. I have already mentioned the work of Russell and Whitehead. Their attempt in the *Principia* to make symbolic logic the foundation of mathematics was part of a general effort to realize Leibniz's goal of reshaping language into a suitably logical tool for philosophical inquiry. "Logic is no longer merely one philosophical discipline among others," wrote the positivist Rudolf Carnap under the spell of the *Principia*, "but we are able to say outright: Logic is the method of philosophizing" ("The Elimination of Metaphysics through Logical Analysis of Language," in *Logical Positivism*, ed. A. J. Ayer, 133). The logical positivists believed that the analysis of language was the only legitimate task of the philosopher; for them, the old philosophy, metaphysics, was literal nonsense because any metaphysical statement used words in an illegitimate way and so made no sense. They further believed that what could not be said could not be thought, and in criticizing essays on metaphysics, they never addressed the arguments

put forward but simply analyzed individual sentences and showed
them up as vacuous—an approach calculated to infuriate their
opponents. The positivists, then, reversed the project of earlier
proponents of a universal character. Instead of extending mathematics to cover a wider domain of experience described by ordinary language, they proposed to reduce ordinary language
to the status of a prose commentary upon the equations of the
physicists.

Today the positivists have given way to a dozen philosophical
sects. Their scientific messianism has been abandoned, but their
emphasis on the logical analysis of language has been retained
and refined. The positivist creed is still very much alive: "It is the
peculiar business of philosophy to ascertain and make clear the
meaning of statements and questions" (Moritz Schlick, "Positivism and Realism," in *Logical Positivism*, ed. A. J. Ayer, 86).

Silent Structures

In the dichotomy between the poetic and logical view of language, the logical has held the initiative now for centuries, as we
have developed steadily away from the oral culture of Greece and
Rome and medieval Europe. This is emphatically not to say that
we have not had great poets but rather that ours is not an age of
poetry. Even in the Middle Ages, there were some (the nominal-
ist philosophers) who argued that names were purely matters of
convention. That argument became much more persuasive after
the invention of printing and the mathematical revolution had led
to silent reading and the manipulation of symbols assigned by
convention. With the advent of the computer, logic continues to
triumph over poetry. The whole course of linguistic philosophy
from Leibniz to the positivists seems to culminate in the computer, where symbols are drained of connotations and given
meaning solely by initial definition and by syntactic relations to
other symbols.

The positivist creed itself belongs to the age of print, not the
computer age. We may wonder whether earlier linguistic philosophers would be gratified or a bit put off by the extent to which the
computer has realized their vision of a totally logical mode of
communication, for the computer takes their ideas to an extreme
unthinkable in the previous era. With its uniform pages of print,
the press suggested that words were mere tokens to be manipu-

lated by the eyes and the intelligence of the reader. Still, the words did stay fixed on the printed page and did demand considerable mental effort on the reader's part. Reading is work, a fact often forgotten by educators who wonder why children prefer television to books. And the mystery of the mind could not be eliminated from the work of reading as long as the real "word processor" was still the human being who picked up and perused the book.

The press is a rigid mechanism, capable only of repetitions of the same product—pages, the order of whose lines is fixed once and for all by the printer. But the computer, which can change its logical shape and purpose as it runs, escapes the rigidity of mechanical word production. Words no longer stay put on paper; they thread their way through the memory and processing unit. No longer is a human intelligence required to manipulate verbal symbols; the computer does so according to a program, such as a compiler which reads and translates statements in FORTRAN without any human intervention. Now there are no mysterious mental processes at work, as there always were with the book. The computer's elimination of mind from the act of reading is a startling change. As long as there were minds involved in the "processing" of a language, it was difficult to treat language entirely as a collection of arbitrary symbols. Human memory is by nature resonant, setting up word associations that defy logic, giving connotations to words beyond their definitions, drawing analogies between the outside world and the words that represent it. With words and symbols fixed on the page, even the logician had to rely on his memory and his faculty of reason to set the symbols in motion. Now, when the computer puts them in motion for him, there is no possibility of resonance or analogy interfering with the rules of logic.

Computer language is, as the logicians had hoped, the triumph of structure over content; to be more precise, it is a reinterpretation of content (what linguists call "semantics") in terms of structure. Never before has language anywhere outside of the logician's study possessed such structural clarity, such purity of form. And never before has the logical view of language found application in such a variety of practical endeavors. In one sense, each new problem programmed into the machine—each new compiler, information retrieval system for business or science, secretarial word processor—is a further conquest for the logical view of language.

It is no accident that in our age linguists are coming to regard human language very much as computer specialists regard their codes. We began by emphasizing the differences between natural language and computer codes, but to many the similarities now seem more significant. Modern linguistics is by no means an offspring of the computer; rather, both linguistics and computer language are children of their day, working synergetically to change the culture that gave birth to them. The work began with the structural linguists of the forties and fifties, who analyzed English hierarchically (from words to phrases to clauses) and by mechanical procedures they hoped would eliminate altogether the question of meaning. Then Noam Chomsky's book *Syntactic Structures* appeared in 1957 (the same year as the release of FORTRAN), with its proposal for "transformational-generative" grammar. Chomsky's approach and others like it have been vastly influential in the English-speaking world.

The new approach is to treat language as an algebraic structure rather than as a lexicon of individual words. A noun in isolation is of little interest; what matters is the way in which nouns, verbs, and other parts of speech may be joined together to "generate" sentences. The trick is to identify rules that allow the production of legitimate sentences and to use these rules to describe as much of recognized English as possible. These rules are often written in abstract symbols and sometimes closely resemble symbolic logic. One popular scheme is to assign a tree structure to every sentence. At the top one writes a single symbol, S, for the whole sentence. At each lower level the structure is described in more detail until the words of the sentence themselves appear as the leaves of the tree. Each level is generated from the one above by such rules as S := NP VP, which indicates that the symbol S has two offspring, NP and VP, and means in old-fashioned grammatical terms that every sentence (S) consists of a subject (noun phrase = NP) and a predicate (verb phrase = VP). This grammar is generative because one can use its rules to generate sentences (or in reverse to parse them). Equally important is the idea that one grammatical structure may be transformed into another without a change of meaning. Thus, a sentence in the passive voice, "John was electrocuted by the central processor," is a transformation of the active voice, "The central processor electrocuted John."

What is interesting is how closely these algebraic manipulations of natural language agree with the computer's manipulations of its codes. Linguists draw tree diagrams to analyze English sentences, just as compilers create them to process statements in FORTRAN. Like FORTRAN, the linguist's English is a string of symbols, each lacking content until defined and each standing in an ordered relation to other symbols in the string. As with FORTRAN, the processing of English is the transformation from one form (a string of words) to another (a tree diagram) or back again, according to a series of rules of replacement. Computers transform input statements into output, just as the linguist's grammars transform the active voice into the passive. Computer specialists now readily acknowledge the influence of recent linguistics in their practical and theoretical efforts. Conversely, their computer languages are the only ones that are perfectly described by such generative rules; complex and irregular natural languages have proven less tractable to the algebraic approach.

Simple exercises in generating English can be programmed into the computer. In fact, such programs have become a standard assignment for beginners. The program is given a vocabulary as well as simple but legitimate structures for combining vocabulary items into sentences. It then casts words together by random selection. In the simplest version of the program, no restrictions of meaning are imposed: the program simply inserts any noun in a place appropriate for a subject or object, any verb in a slot requiring action, and so on. The resulting sentences are syntactically perfect and often ridiculous (see figure 8-2). They are also, if read one after another, strangely unsettling. One begins to seek rather desperately for meaning behind the syntactic patterns, and one's laughter becomes almost nervous, as if a machine, allowed to say anything at random, might actually say something profound, even embarrassing.

That brings us back to the question of meaning. Questions of language soon become questions of knowledge: how language helps us attain knowledge or, if you like, where the meaning lies in or behind our words. Chomsky himself, fully aware of the larger issues that lurk behind his linguistics, has said that language is "a mirror of the mind"; he has gone so far as to define the mind as "the innate capacity to form cognitive structures," structures that are "represented in some still-unknown way in the brain" (*Reflections on Language*, 4, 23). A colleague of Chomsky, Jerry Fodor, has elaborated this idea in a book entitled *The Language of Thought*: that language is of course not English or German but rather some kind of internal code wired into our

Figure 8-2. Computerized English

Production rules for the automatic generation of English sentences
(":=" indicates that the symbol on the left is replaced by the string
on the right, and "|" specifies alternate versions of the same
rule):

```
S    := NP VP
NP   := DET ADJ NOUN | DEM NOUN
VP   := ADV VERB NP | VERB NP | VERB NP PP

NOUN := Pierre | lamp | Andre | Natascha | Anna | castle | carriage
VERB := kissed | seduced | kicked | abandoned | married
DET  := the
DEM  := this | that
ADJ  := loquacious | ill-fated | spiteful | irate | hopeful | hopeless
ADV  := seldom | never
PREP := from | with | in | without
```

A tree representation of one sentence generated from these rules:

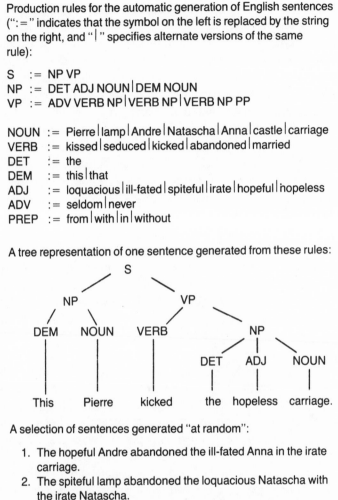

A selection of sentences generated "at random":

1. The hopeful Andre abandoned the ill-fated Anna in the irate carriage.
2. The spiteful lamp abandoned the loquacious Natascha with the irate Natascha.
3. The irate Natascha married the hopeless carriage over the spiteful Anna.
4. The irate Anna seduced the spiteful lamp with the ill-fated carriage.
5. This Andre married the ill-fated lamp.
6. This Pierre kicked the hopeless carriage.
7. That castle kissed the hopeful Anna.
8. The loquacious Natascha married the loquacious Natascha.

The sentences obey the structural rules of English, but they are
often senseless. They illustrate the computer's ability to separate
structure from meaning.

brains at birth, a code expressly compared to a computer's machine language. In thinking, we transform messages in this code
from initial states through intermediate forms to output, which
may be verbal or muscular.

A little thought, a few transformations of our own, will show
how far we have come from the idea of meaning common to earlier cultures. Before the watershed era of printing and mathematical physics, and indeed well into that era, the sound, the
vivid presence, of individual names and words meant something
by itself. The emphasis was on nouns as names, whereas the relationship among words in a sentence was largely a matter of how
their meanings would collide and combine. But in the age of the
computer, silent, spatial structures are used to map out the meaning of language. And the meaning of a sentence is its structure,
which our mind, like a computer, can represent, distill, and transform. For the mind itself is the capacity to form structures. Linguists have by no means abandoned questions of semantics; they
now refer to the "lexicon" that human beings carry around in
their heads. But in fact, words in this lexicon are defined by the
relationships that they permit. "Dog" is defined as a word to
which certain classifiers apply (animate, animal, domestic, and
so on); it may serve as the subject of such verbs as "bite," "run,"
or "bark" but not of "refute," "applaud," or "vote."

Some prefer to define "semantic networks," in which the
nodes stand for concepts, events, persons, or objects and the
links stand for associations among these concepts. This definition
too converts the human experience of meaning into a spatial
structure, which can be graphed on paper or stored in a computer
memory. The symbols placed at the nodes are essentially empty,
without resonance or depth, marks on paper or bits in the machine. Nothing can be done with such symbols except to trace
out, add, or delete links among them. Any less tangible notion of
meaning is itself meaningless for the computer.

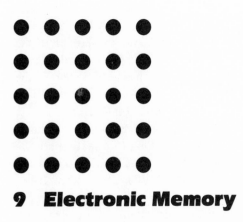

9 Electronic Memory

The central processing unit of a computer is of no use without something to process, without programs to direct its high-speed manipulations and data to manipulate. The part of the machine in which elements of programs and data await their turn in the logical grinding mill (the CPU) and to which they return after processing is called the computer's "memory." Memory is what preserves instantaneous calculations, so that they can be used in further calculations or be written to a peripheral device for humans to read and use. Every instruction and bit of data that passes through the central processing unit comes from and returns to memory. A computer without a sizable memory is merely a calculator, accepting data, performing the arithmetic, and handing the results back to a human operator. It is no more powerful or culturally suggestive than the simplest pocket calculator, which does embody numbers but does not seem to embody logical thought because it depends directly upon the man or woman who pushes the keys. Substantial memory frees the computer from immediate human control. The machine can move under its own logical steam and accomplish long, seemingly human feats of calculation "by itself."

Electronic memory includes in the largest sense all the devices and programs that store and retrieve information. To speak of these logic machines and algorithms as "memory" is an implied comparison to the human act of memory. The comparison comes naturally, for electronic technology is so alive and flexible that it seems to many not merely to rival human memory but to explain it. The urge to understand human memory by reference to the computer's storage devices has proven irresistible to modern psychology. In that sense, computer memory has become part of the larger metaphor of the electronic brain, and this aspect will be considered later in the discussion of artificial intelligence.

Here I simply want to explain computer memory as a technology and to show that storing and retrieving information are the latest developments in a long technical tradition, to which ancient, medieval, and modern cultures have made their contributions. "Storing and retrieving information" is computer jargon, but every age has needed to possess, maintain, and render available its fund of experience and knowledge. Memorization, writing, and printing were all previous memory technologies, and as always, computer memory shares and rejects qualities of each.

Digital Memory Technology

An electronic memory device is any machine or component that fixes the evanescent signals of the central processor. There is first of all the memory unit of the von Neumann computer, which contains the storage cells directly accessible to the CPU; this unit is sometimes called "internal memory." It is here that programmers create their geometric structures of data. This internal memory is discrete and consists of a string of storage locations, each holding one unit of data (one or a few bytes). Each location has its own address, so that it can be used individually by the processor for reading old data and writing in new. Internal memory allows for *random access*; the central processor can proceed directly and quickly to location 1000 without examining locations 1 to 999.

Such memories must be built from relatively expensive components. Today engineers use transistors organized into minute integrated circuits etched onto wafers of silicon. Integrated circuit memories are becoming more compact and so more capacious each year, but expense still limits the amount of this very fast, random-access memory with which a computer can be equipped. Although it is reasonable to store a few, sizable programs in a computer memory at one time, it is seldom desirable to store, say, a coded version of the entire contents of a book. But computer programs often need large amounts of data, and they often produce mountains of output. For these purposes, engineers have created the mass storage devices that form the system's "external memory."

Each external memory device has its own operational characteristics, and these are of vital concern to engineers who design systems and programmers who use them. The machine most familiar to laymen is the tape drive, which functions like a home tape recorder but at much higher speeds and with simpler data. Until recently, the whirl of tapes was a favorite technique of filmmakers to show the computer at work, for the tape moves visibly and quickly, whereas the electrons that perform the calculations are not accessible to the camera. (Now, by the way, graphic displays on television terminals seem to be the most popular image.) As the reel of tape speeds through the drive, the drive reads or writes by electromagnetic induction patches that correspond to binary digits. The result is a compact and lasting record. Storage and retrieval of this record is fast while the tape is spinning, but the tape is not always spinning. It may have to be removed and replaced by another tape if a different record is needed. Mobility is the key: tapes can be carried from one machine or computer center to another, and any one tape drive can service an indefinite number of tapes. However, if the computer needs data on a particular tape, it may have to wait minutes or hours for a human operator to find and mount the reel. Nor is the tape drive a random-access device. Even when the proper reel is mounted, the processor cannot simply obtain the 10,000th byte of information. It does not know exactly where that byte is; it knows only that it comes after 9999 other bytes from the beginning of the reel. It must count its way through from the beginning, just as we would leaf through a book to find a passage if there were no page numbers. The drive allows only *linear access*, and retrieval may still be a matter of seconds rather than the millionths or billionths of a second by which random internal memory is measured.

I could go on to describe such past, present, and future storage mechanisms as the acoustic delay line, the disk drive, and magnetic bubble memory. Each brings special qualities to the tasks of storing and retrieving data. Each has or had advantages and drawbacks that make it the best solution for some storage needs. But the point is simply this: the computer specialist analyzes memory devices in terms of the medium of storage, the density (amount per unit of volume) of data stored, the speed and ease of retrieval, and the cost. The important question is not what is stored, for the answer is always the same—binary digits. The question is how these bits are to be stored and retrieved in a device that is as capacious and as fast as his budget will allow.

Computer memory is software as well as hardware, programs to put the core storage and the external devices to use. The issues are all those we have encountered before: how to represent the data, how to manipulate those representations, and how to operate within the space restrictions of the machine. Internal memory is fast but expensive and thus limited in size; external memory is relatively slow but plentiful and less expensive. The trick is to strike a balance between the two, to use internal memory as efficiently as possible and to resort to external memory as seldom as possible. It is another version of the problem that confronts the computer specialist at every turn, the husbanding of scarce resources, and this problem is solved again by structure and standardization.

Recall the hierarchy by which the programmer organizes his storage space and the hierarchy of languages he uses to command the processor. Memory is also organized hierarchically. The storage features built into computer memory become building blocks for the further structures of the programmers. Bits, bytes, words, and pages are usually part of the computer architecture, its initial design; the systems programmers then group pages into segments, segments into files, files into libraries—thus building up larger and larger units of information that allow the users of the computer to organize their data. A user may have several libraries—one for programs, one for data, one for personal correspondence—each with its own identifying name. Within a library are separate files, again named, one for each program or personal letter. The casual user may never need to delve beneath the level of files; he is protected by layers of control programs from the details of where and how the files are stored. Indeed, he may even be able to ignore the very distinction between internal and external memory if the control programs, written once for all, handle invisibly the problems of moving data from one device to another.

At the top of this hierarchy of hardware and programs are the so-called *data base systems* for organizing large amounts of information. The name suggests something solid and reliable, and indeed business does rely upon these systems for such practical purposes as bookkeeping and inventory; libraries use them as well for cataloging and searching bibliographies; there also exist data bases specifically for medical journals and for law reports.

Since these systems can run to millions of entries, sheer volume
requires them to be ruthlessly hierarchical in their organization. The process of selection and "memorization" is carried out by dozens or perhaps hundreds of catalogers and typists, who punch customer invoices or book titles into the machine.

A data base may consist of any kind of discrete information put into "machine-readable" form. The point of storing the information is to get it out again, easily and quickly. So it is normal to analyze the data upon entry, to break it into convenient fragments and then organize the fragments for easy retrieval. There is no apparent way to atomize, say, the text of Shakespeare, but most functioning data bases deal with inventory, billing, or bibliographies, not with literature. Let me illustrate with a data base that corporate America is most unlikely to think of generating: a data base of the emperors of Rome from Augustus to Romulus Augustulus. In our organization, each emperor would have one record of data, and the records would be stored in chronological order. In addition to his name, we might include the following: when he reigned, the probable cause of death, and (just for fun) his principal vice according to Roman tradition. We can represent this information as a table (figure 9-1).

How would this (somewhat prejudicial) information actually be stored? Each record would be given a location in internal or external memory, and that location would have an address associated with it. The address is the link between this record and the others in the system. The address makes it possible to build structures out of the items of data, perhaps trees or networks of various kinds. The individual fields within the records may enter into structures of their own. Such structures, built like tinker-toy houses out of records and links, are familiar from the discussions of space and language. They are possible because the information about each emperor has been atomized: what we have is not an organic account of the life of each man but instead a set of discrete facts about each. Indeed, the need to fit the information into discrete categories forces us to be rigid and arbitrary. This is true in business data bases as well; for any category, there are exceptions that must be distorted or dropped from consideration.

What of retrieval? The kinds of queries that can be made of the data base depend upon the structure of the data and the sophistication of the query programs. The programs often rely upon set theory to gather up the needed records. The user demands: "Give all the emperors whose names begin with the letter A" or "List

Figure 9-1. Data Base of the Roman Emperors

Name	Reign	Cause of Death	Principal Vice
1. Augustus	31BC–AD14	natural	superstition
2. Tiberius	14–37	natural?	lechery
3. Caligula	37–41	murdered	psychosis
4. Claudius	41–54	natural?	indecision
5. Nero	54–68	suicide	psychosis
6. Galba	68–69	murdered	indecision
7. Otho	69	suicide	vanity
8. Vitellius	69	murdered	gluttony
9. Vespasian	69–79	natural	avarice
⋮	⋮	⋮	⋮

the emperors who reigned in the year 69." More interestingly, a question may combine values from several fields of the record: "List all the emperors after Vespasian who died a natural death" or "Give all the emperors from the second century whose principal vice was fratricide and who did *not* die a natural death." The programs will follow out links and identify records with just the right combination of fields. The above queries have been written freely in English, but commercial data bases may require some special code words or rigid syntax. The user cannot ask such a data base as ours for any more enlightening or elaborate answer.

A data base is really a memory in the strict sense: the only thing that comes out are the records that were read. Queries may, however, reveal structural relationships among the records that the human eye could not easily perceive. Relationships are always expressed by the retrieval of an answer set, a number of records that have all the intersecting qualifications we have specified. Of course, our example is tiny compared to an application in business or library work. Often the sheer mass of information makes storage and retrieval by other, more flexible means impossible.

It may seem that the concept of computer memory has little to do with the Western or ancient history of the idea. The capacity of a computer to store and retrieve data depends upon mid-twentieth-century technology—the invention of fully automated writing pads upon which a processor can engrave electronic messages and later read them back—and no previous age possessed such a technology. Nevertheless, the ancient and earlier European views of memory do deserve our attention for technological reasons. Memory technologies (even in the computer's sense of information storage and retrieval) existed in both cultures: techniques of memorization and writing on papyrus rolls in the ancient world, manuscripts in the Middle Ages, and finally printing. Electronic memory has something in common with all these earlier technologies.

In Greek mythology, Memory was the mother of the Muses. More was meant by this than merely that a poet must remember a song or poem in order to perform it. Poetry itself was an act of reminiscence designed to elicit a response in the memories of the audience. Greek epic poetry and even early history had as its goal to preserve in the minds of contemporary men the legendary deeds of their ancestors. Greek epic seems to have flourished before the art of writing was known, as an oral tradition passed from one generation of poets to another. With deliberate anachronism, we could say that this oral tradition, found in many illiterate cultures throughout history, is the first human form of information storage and retrieval. It is highly associative and imaginative: the poet evokes for his audience a whole world of epic heroes and gods and fixes in their memory a heroic ideal of conduct for men and gods.

The art of writing, imported from Phoenicia, did not destroy the need in the ancient world for an accurate and extensive memory. Books in the form of rolls of papyrus were cumbersome and difficult to reproduce; they were not designed for reference because it was hard to disinter a passage of prose or a line of poetry from the surrounding material on one roll. Clearly, the reader had to remember the gist of works and perhaps exact quotations as he read if he wanted to make later use of the material. This problem came in addition to the problem of finding a scarce book in the first place. As we noted in the chapter on language, books in Greek society were rather an aid to an individual's memory than a

replacement for it, and the ancients continued to regard memory
as the key to making knowledge useful. Plato was not the first,
but he was perhaps the most forceful exponent of the idea that
all learning is simply a process of memory. In his dialogue the
Phaedo, memory was seen as the tool by which men gain access
to knowledge of their former lives, the contact point between the
individual identity and the eternal soul. As we have seen, Plato
himself was so impressed by the power of pure memory that he
criticized the invention of writing for destroying men's incentive
to remember.

Even after the development of the technology of writing, the
ancient culture remained an oral culture. The continued emphasis
on the spoken word itself put a premium on memory. One wanted
to carry the words of a poem or speech around in one's head, to
have them always available, always resonant. An oral culture fosters a kind of associative thinking that is less common in a culture
of print. For human memory is emphatically not linear, not
bound to recalling ideas or events in a fixed sequence: it can recall events in any of a variety of orders, based upon one shared
quality or another. We do not have to remember an entire play
from beginning to end; our minds leap magically, or at least not
mechanically, to a particular scene or speech that has made a
vivid impression; they then move back and forth through scenes
and speeches that are related, provided, of course, that we have
trained our memories to retain the play, as the Greeks had clearly
done.

So important was memory to the ancient world that it was can-
onized as one of the five rhetorical virtues. In addition to simple
memorization, which was fostered by constant practice in the
theater or the law court, the ancients developed an elaborate art
of memory, to be used particularly in giving speeches. The trick
of this artificial memory was to employ striking images to retain
ideas in the proper order. The student would fix in his mind a
large building, for example, a temple or a Roman villa with a
number of rooms and particular places within each room. He
would then associate striking images with the various places, so
that the images became visual representations, hieroglyphs, of
the points to be remembered. If the introduction of his speech
was to emphasize the treachery of his client's enemy, the student
imagined a man holding a dagger and placed this image at the
doorway of his mental villa. In the hallway he placed images to
remind himself of the details of the charges against his client—

a sick, aged man in bed holding a pair of wax tablets would represent the will of his client's father. In the inner garden of the villa, he placed images to recall his refutation of the charge. And so he proceeded to the impassioned peroration. When he actually delivered the speech, he traversed the house once again, from the vestibule to the inner rooms, stopping before each of the images he had created long enough to deliver the sentence the image represents. We can imagine how our data base of Roman emperors could be committed to memory by this method, with vivid images for the manner of death and the vicious habits of each emperor.

This peculiar ancient practice was in fact a method of information storage and retrieval in the modern sense. It was technical, methodical, and mechanical in application, although the use of colorful imagery made this art of memory more charming than a modern data base. At any rate, the art survived the destruction of the Roman empire in the West and was perhaps as popular as ever during the Middle Ages and the Renaissance. Frances Yates has charted the whole development in *The Art of Memory* and shown us how the mnemonic technique changed its function to suit the temperament of the day. The Greek or Roman orator who practiced the art needed to remember speeches or complicated legal issues. When artificial memory was taken over by such medieval thinkers as Albertus Magnus and Thomas Aquinas, it was given a moral dimension. The monk or scholastic philosopher did not need to memorize judicial speeches, but he did need constantly to recall and meditate upon man's position in a hierarchically ordered, religiously charged universe—to remember the joys of paradise and the torments of hell, the various virtues and sins.

In the Renaissance (and even before in the case of the magician-theologian Raymond Lull), memory images again took on a new meaning at the hands of men who combined diverse influences of the Hermetic and Jewish cabalistic traditions to create their own brand of Renaissance magic. Among these was Giulio Camillo, born about 1480, who devoted his life to the building of a "memory theater." This was apparently a small closet in which two men could stand; it contained a stage and seven tiers of seats with seven gangways. At each intersection between the gangways and the tiers, there was a box with an identifying image, a highly resonant symbol possessing magical as well as historic and philosophic significance. Behind each image there was a drawer holding written materials, such as speeches by Cicero on topics represented by the image. A candid computer specialist would have to admit that the memory theater of Camillo was the first data base, not merely the idea but the physical realization of such a system.

Giordano Bruno, the greatest Renaissance magus, wrote several treatises on memory, working along the same lines as Camillo. He too meant to use artificial memory for establishing a sympathetic rhythm with and control over the universe, but he preferred concentric wheels to Camillo's theatrical metaphor. His scheme began in the center with "star images" and radiated out to a huge wheel with the names of 150 inventors of the human arts and sciences. Such an "astral power station" (Yates's phrase) with its interlocking wheels was clearly a fanciful extension of the contemporary wheel-and-gear technology of the later Middle Ages and the Renaissance. The memory system of Bruno moved while the magus remained still and contemplated or supervised the movement—just the reverse of the ancient system, in which the orator walks through a house filled with static images. Bruno lived in the age of the mechanical clock.

The other memory techniques of the ancient world also continued in the Middle Ages and into the Renaissance: simple memorization (without appeals to magic) and the manuscript. The great change came with the introduction of the printing press. Writing alone had not destroyed the need to remember because throughout the Middle Ages, as in the ancient world, manuscripts remained expensive and hard to reproduce, with each one unique. Printed books, however, were easier to read and exactly reproducible in large quantities. Standardization led to the widespread use of indexes and tables of contents, devices for finding a particular passage or idea. These advantages were not lost on scholars of the day. As soon as the press arrived, their great concern was to turn manuscripts into printed editions; in contrast to today's change, scholars welcomed the change from manuscript to print. In the long run, printing doomed the art of memory. It is true that Giordano Bruno and other Renaissance magi wrote copiously for publication. But a magical text was more appropriate as a decorative, time-worn manuscript than as a clean, clearly printed book, partly because the manuscript conveyed more effectively the impression of a secret being grudgingly shared. More important, the magic words of Bruno's memory wheels surely had to be spoken aloud. Magic in any age requires incantation and implies a view of language that belongs to oral cultures rather than a culture of printing and silent reading.

As a device for storing information, the book assured perfect but unalterable copies. A sentence that was set in lead type, copied in ink thousands of times, and sent throughout Europe or indeed the world had an excellent chance of surviving unchanged for hundreds of years in the hands of thousands of readers. The situation was far different with earlier manuscripts; errors by copyists and copious marginal notes, added perhaps by several owners over several centuries, made every manuscript unique. One scholar would borrow another's copy of the same text in order to collate and improve his own. The manuscript was actually more flexible than the book that superseded it. I do not wish to overstate the case: a printed page developed in time into a marvelously subtle medium for ideas, with its varieties of type fonts and inks, its pagination, tables, indexes, bibliographies, and all the rest. But, however subtle, the book is a rigid medium: printing the copies (storing the information) is carried out only once at a central location, and changing the plates is hard.

In its rigidity, the book was characteristic of the whole mechanical era. For just as ideas were fixed on the printed page, so mechanical information was fixed in the gears of the clock, the steam engine, or the dynamo. The mechanical-dynamic age was populated by machines that did one or a few things superbly well. Their range of activities was generally more limited than the simpler devices they replaced, and anything more than small adjustments might require rebuilding the machine. Technologists of course understood the value of flexibility and sought to build it into their machines as far as their stubborn materials would allow—hence the variety of self-regulating mechanisms in watches, engines, and the like.

Gradually, ways were found to store and express information mechanically. Perhaps the best example was the Jacquard loom, invented around 1800 to weave patterns in silk automatically. The loom was controlled by a series of wooden cards, in which punched holes determined the raising of various warp threads to generate the pattern. The pattern could be altered by simply punching new cards. This solution caught on. Babbage intended to use punched cards for his Analytical Engine, and Herman Hollerith at the end of the century successfully used them in tabulating the American census. Then began the great era of tabulating machines, which read, collated, duplicated, retrieved, and sometimes destroyed endless stacks of cards. The era came to a close in the 1950s with the introduction of computers and magnetic tape. At that time the Social Security Administration was said to have billions of cards, whose information was transferred to tape. Today punched cards are used sometimes for the input of programs and small amounts of data, but the manipulation of the data is usually wholly electronic.

Mechanical tabulators for business data were ingeniously flexible in their use of an inflexible medium. Each card could be punched once; any mistakes in the punching meant that the card must be thrown out. Data on cards could be manipulated only by moving the cards physically around in the machine from one bin to another. Again, our data base of Roman emperors could have been entered on punched cards; queries would be handled by running the whole deck through the tabulator (perhaps several times) to select exactly those cards whose fields satisfied the conditions of the query. Business processing by tabulators in fact has much in common with business processing by computers (it is not surprising that IBM made the transition so successfully from selling one kind of machine to selling the other). But the crucial capacity for split-second decisions, for the working and reworking of the data, for the building and demolishing of structures of information—all this is simply absent. Probably no one ever talked of replacing the book with card printers and punches, as people often suggest that the computer will make the book obsolete as a memory device.

Information Retrieval and Electronic Power

The computer's technology of memory fits into a tradition at least as old as Greek civilization. It is in one sense a triumphant extrapolation of the mechanical technology that preceded it. In another sense, its flexibility recalls an earlier era, before the book or the tabulating machine.

Like the book, computer storage is capable of faultless representation and near perfect reproduction. The assembly line is something we associate with the mechanical age, and yet the production of identical units of information is even easier with the computer than it was with the printing press. The computer goes the press one better. The book is a carefully designed structure of words on a page, but it is a frozen structure, and it is linear. Most

printed books have a sense of direction and development; they
expect the reader to follow the flow of events and ideas described from beginning to end. As I have stressed, elements of data in a computer memory system are not limited to one rigid order. Put the elements in a random-access device, and the user can examine them in any order he cares to define. With the Roman emperors, we are not confined to listing the records in chronological order. We may jump around by looking at other fields, such as principal vice or cause of death; we may retrieve the same records several times in several different orders. In short, because we define and redefine the structure of our data, we break free of the fundamentally linear order imposed by the mechanical technology of the book. Our structures have two or more dimensions, as is indicated by the multidimensional trees and other diagrams we draw to represent them. The computer memory offers us in a strict sense a "new dimension" in the representation of information.

In building such structures, computer memory is associative rather than linear. It allows us to follow out networks of association in our data, as indeed human memory does. This fact was particularly appreciated in the oral cultures of Greece and Rome, the Middle Ages, and the early Renaissance. The computer also resembles the manuscript, and human memory, in its capacity for change. It is always ready to replace old information with new. Anyone who uses the computer knows the ease with which even valuable information can be inadvertently "written over" and destroyed. The manuscript was easier to correct than the book, and its margins were always available for glosses or additions to the original text. As for human memory, it is perhaps not so easy to erase, but one can add and modify memories almost indefinitely.

The printed book is often said to have ruined the Western memory. With such a convenient device for preserving knowledge, men and women hardly needed to exercise their "internal storage" any longer. As books and other written aids became common, people lost the ancient Greek or even the Renaissance capacity to retain a play, a poem, a speech as they heard it. Perhaps we could put this another way: the book was such an effective solution to the problem of memory that it killed interest in the subject. If so, electronic memory technology will mean a rekindling of interest. Memory is important to Turing's man simply because so much of the computer's operations depends upon techniques of storing and retrieving data. It is probably true that all the information being generated by today's society cannot be mastered except by electronic methods. Mastery means subsuming that information into data structures, storing it in data bases, and retrieving it silently and efficiently on television screens, in other words, getting the information under control and preserving it in such a way that it can be found again.

Such mastery is probably as important to the computer specialist today as it was to the Renaissance magus; the two share the feeling that memory is the key to human knowledge and therefore to human control of the world. The memory devices of Camillo and Bruno were meant to reflect the true structure of the world, not merely or principally the world of sense experience but all the realms of humanly accessible knowledge. To be able properly to manipulate the devices was to win that knowledge and, through sympathetic magic, to establish a rhythm between oneself and the world. The computer specialist, of course, does not believe that his programs have supernatural powers. He does regard his memory systems, both programs and hardware, as attempts to reflect the logical structure of the world outside the computer.

Computer memories are also powered machines, although they substitute electricity for the more dubious astral influences that animated Renaissance memory devices. And if a computer memory is capable of less spectacular tricks than those of Camillo's theater, its tricks are more useful and more convincing to businessmen and scientists. An electronic data base is the modern, rationalized equivalent of a memory wheel: its purpose is to allow the user control over an amount of information far larger than he could keep in his head. The information in current data bases may be bibliographies of medical journals, but computer specialists are already thinking on a grander scale. In the 1960s one computer firm commissioned a study of the feasibility of storing all recorded human knowledge in one giant data base. If it is ever attempted in the coming decades, such a data base will not only be the heir to the French *Encyclopédie* of the eighteenth century (itself a product of enlightened empiricism) but also the heir to Camillo's memory theater of the sixteenth century. It will be an electronic embodiment of the conviction that memory is knowledge and knowledge power.

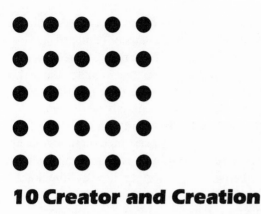

10 Creator and Creation

This chapter is about making things electronically: the relationship of the computer specialist to his materials. We have already discussed the principal hardware and software components of the digital computer—the central processor, the core memory and external storage devices, the timing mechanism, and the programming languages. The machine is complete. In order to run, it needs one final element, the programmer. How does he fit into this intricate technological scheme? How does he impose his sense of purpose upon the otherwise purposeless, blank page of the Turing machine?

The reader may feel that creativity is the wrong word to describe what the programmer does. The word used to call up visions of the artist struggling to embody great truths in intractable marble; today it is applied, at least in America, to every act of self-expression, no matter how trivial. Computer specialists themselves are fond of comparing elegant programming to the work of an artist or poet. But I suspect they have in mind the academic painter or poet, the artist who produces something polished and elegant while working within narrowly defined rules of how the human body should be drawn or what language is appropriate to the emotion of love. I do not think that many programmers would want to compare their results to the paintings of Jackson Pollock or to see themselves as Romantic poets, inspired interpreters of some higher reality. At any rate, computer special-

ists are in spirit and practice craftsmen rather than artists, at least
if there is a meaningful distinction between craft and art.

Still, technical invention does deserve the name of creativity. It
has certainly been a part of culture longer than high art; the making of tools by hominids long antedates the cave painting in the
Dordogne. Making tools and shaping materials were mankind's
first creative acts, and the desire to make things fit together effectively and attractively has been with us ever since, in more recent
times joined by the desire to make things work in a mechanical
sense. This kind of creativity is important to a culture intellectually as well as materially. How we think about the act of
creation is at least in part determined by our current tools for
technological creativity. Indeed, what we can make in our own
technological world influences our beliefs about the world of
nature and how it came to be: hence the ambiguity of the word
"creation," which refers to both the act of making and the world
as a product of (divine) creation. Here again, technology and philosophy interact. The computer, the most philosophical of machines with its preoccupations with logic, time, space, and
language, suggests a new view of human craftsmanship and creativity as well.

Coherence and Correspondence

It is the programmer who chooses what problem to set for his machine. Like the engineers who build computers, the programmer
has the character of a professional technologist and often works
as a member of a team. Good technical programming for creating
new languages, control programs, or programming tools may require years of training and a mastery of mathematics, if not solid-
state physics. Such software projects may involve dozens of
programmers; here team spirit is as important as in any other
technical field. But programmers need not always be specialists,
and they need not make anonymous contributions to a group
effort. Nearly anyone can learn to use a modern computer system
and can then work alone on all sorts of problems, mathematical
or symbolic. High-level languages and programming "packages"
allow even the beginner to exploit much of the power of the machine. And *multiprogramming* techniques allow each user, naive
or professional, to work in near perfect isolation from the others
in the same system.

To a programmer seated at a terminal, it appears that the entire
system (central processors, tapes, disks, printers) exist solely for his use—a vast electronic sheet of paper upon which he may write almost undisturbed. With the help of a remote terminal connected to the processor by telephone lines, the programmer may in fact be alone in an office or at home, miles from the machine. Anyone who has worked in this fashion, alone, late at night (the ideal time for programming because there is less demand for the facilities, which are therefore more responsive and reliable), knows the peculiar satisfaction of this hermetically sealed kind of creativity. The feeling is like that produced by lone work of any kind, which men have experienced since the beginning of civilization: plowing a field, puzzling out a theorem of geometry with straightedge and compass, constructing a watch.

What especially characterizes the programmer is his withdrawal from nature into the private intellectual world of the program he is writing. Normally, he thinks neither of the keyboard at which he is typing nor of the electrons that are performing the calculations. He concentrates his full attention on the abstract problem, its representation in the programming language, and the logical design of the machine he is using. In this respect, he resembles the mathematician, the philosopher, the theologian, or indeed the chess master, all of whom live more or less completely in intellectual worlds of their own making. By contrast, the farmer is fully involved in nature as he tries to keep his ox or horse or tractor in a straight line and his plough at the proper depth. The clockmaker at his desk is constantly reminded of the intractability of the metals with which he works. Even the computer engineer works from nature, for the most complex large-scale integrated circuits begin as bits of silicon (sand), metal, and other elements. The industrialization of any activity tends to remove men and women from their raw materials and hence from nature. So the farms of modern agribusiness, with their huge and complex machinery, are becoming assembly lines for the manufacture of poultry and produce. Modern spinning and weaving are achieved with human hands touching the thread only to load the machine.

Philosophers and mathematicians have been with us since before Greek times; their abstract labor is familiar. The computer programmer is remarkable because he is the first technological man whose work is divorced from nature in this way. It is the paradox I have stressed from the outset: the abstract process of com-

puting gives pragmatic results. The computer specialist Frederick
Brooks writes: "The programmer, like the poet, works only
slightly removed from pure thought-stuff. He builds his castles in
the air, from air, creating by exertion of the imagination. Few media of creation are so flexible, so easy to polish and rework, so
readily capable of realizing grand conceptual structures. . . . Yet
the program construct, unlike the poet's words, is real in the
sense that it moves and works, producing visible outputs separate
from the construct itself" (*Mythical Man-Month*, 7).

The programmer's task is to set logic to work in the world, and
to do so he must mediate between the problem to be solved and
the rigorous and curiously unnatural brand of logic by which the
computer operates. I present here a simple application of computer logic to a prosaic problem: a program to sort a list of English words into alphabetical order. To command the computer to
alphabetize, one must break the process down into a sequence of
operations that can be performed by the processor. The machine
never understands what it means to alphabetize. However, it can
execute an operational definition; it can be made to alphabetize,
if the process is expressed as an algorithm. The step-by-step algorithm must eventually be written in the symbols of a programming language, but here the steps can just as well be spelled out
in English:

1. Find the smallest (alphabetically earliest) word in the
   list. Take this word from its position in the list and set it
   aside.
2. Move all the unsorted elements of the list down one,
   until the vacant slot is reached, thus emptying the slot at
   the top. Put into this slot the word set aside in step 1.
3. Repeat steps 1 and 2, excluding from consideration the
   portion of the list already sorted in previous repetitions,
   until the whole list is sorted.

In most encoding systems, the letters of the alphabet are assigned values in serial order. Thus the first step is to find the alphabetically earliest name simply by searching for the smallest
binary number. (It is as if the code for the alphabet were the
binary equivalents of the numbers 1 through 26: A=1, B=2,
C=3, . . . , Z=26. The program compares two strings letter by
letter from left to right. "JOHNS" is less than "JONES" because
the first two letters of the strings match, and in the third position

H=8 is less than N=14.) The rest of steps 1 and 2 are instructions
to move data around, to shift strings of bits so that the name can find its proper place at the head of the list. Step 3 is the loop step. After the first execution of 1 and 2, the first element in the list is right, but the other elements are still disordered. The program simply ignores the first element of the previous execution and repeats its procedure on the unalphabetized smaller list. It continues in this fashion until the unalphabetized portion has been reduced to one element, alphabetically the last, and then halts (figure 10-1). The loop could in theory be applied to as large a list as we like: 1000 elements would simply require 999 repetitions of steps 1 and 2.

This procedure seems stubbornly, almost irrationally, systematic compared to the way a file clerk would go about setting a list in order. It is often said that computers operate in idiotically simple ways, that writing a program is like explaining a task to an idiot because one must spell out each step in excruciating detail. It is true that the computer brings to the task outlined above only a handful of relevant capabilities: those of scanning a list, comparing two elements, inserting an element into a list. In fact, these three capabilities are all that are needed to alphabetize, but the price for this simplicity of function is complexity of method. The programmer must describe every step, whereas a clerk sorting a set of file cards need not be so rigid. If he finds three cards with the same last name, he sets them aside immediately. If he finds a man named Mr. Aarkin, he puts him at the front of the file without further consideration. He has a dozen tricks of intuition not available to the computer.

A computer program seems strange to a layman, not merely because it approaches the problem in a simpleminded fashion but also because its approach is dictated by the kinds of operations (simple or not) that computers can perform. An alphabetizing program is not the same as the instructions we would give an idiot because even an idiot would have intuitions that we could put to use. Electronic thinking is defined by a different set of rules than human thinking, at least everyday human thinking. The rules are those of symbolic logic, applied to the symbols stored in the machine's memory. In solving problems, the central processor pulls these symbols out of memory, combines or compares them with other symbols, and restores the result to memory. The processes are far removed from conscious human reasoning in their regu-

Figure 10-1. Sorting Problem

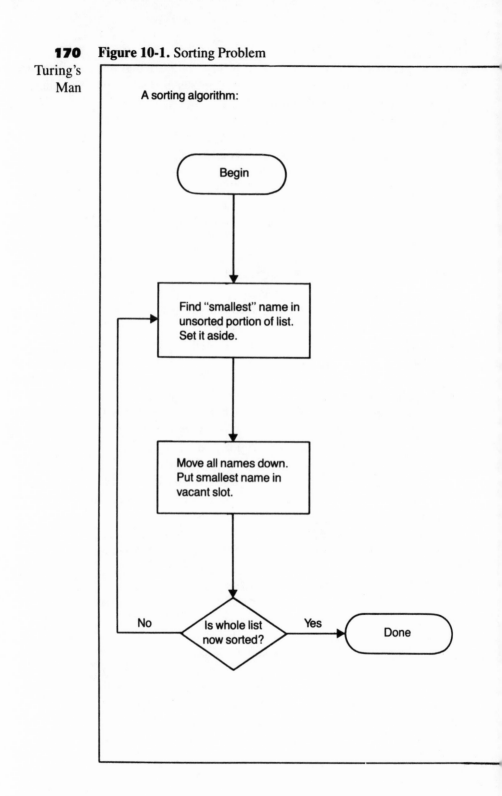

A sorting algorithm:

Begin

Find "smallest" name in
unsorted portion of list.
Set it aside.

Move all names down.
Put smallest name in
vacant slot.

No   Is whole list
now sorted?   Yes   Done

One "iteration" of the loop:

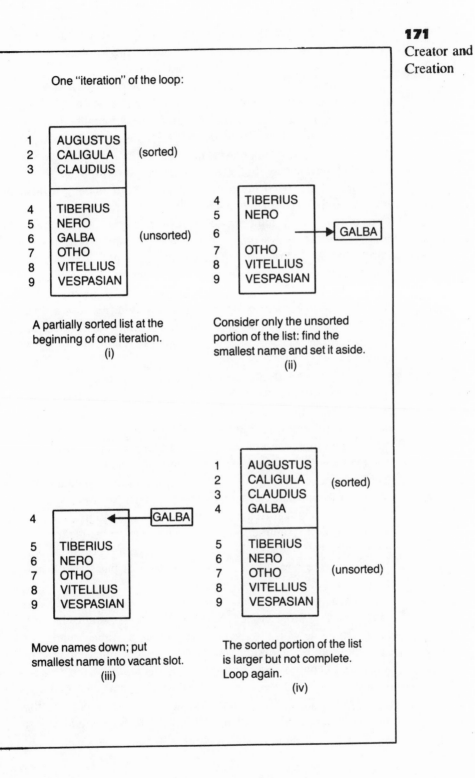

A partially sorted list at the
beginning of one iteration.
(i)

Consider only the unsorted
portion of the list: find the
smallest name and set it aside.
(ii)

Move names down; put
smallest name into vacant slot.
(iii)

The sorted portion of the list
is larger but not complete.
Loop again.
(iv)

larity and repetitiveness, as well as in their abstract structural beauty. Rules of logic concern form, not content; a computer manipulates pure symbols that are devoid of content.

Even our trivial example shows how the logic of a program can be translated into a kind of visual charm. The repetition of lists of words, each list a single step closer to its proper alphabetization, is logic visualized; it represents the relentless progress toward a goal and is all the more striking because not a single step is missed along the way. In the computer's world of thought, shortcuts that might speed the process at the cost of making a false turn are eschewed. A more sophisticated program can create patterns of real beauty, but even our pattern shows how the computer juggles and arranges words, treating them as empty symbols according to its instructions. The computer simply does not approach a problem as we do, even in our most mechanical moments. We may therefore think of programming as the art of transferring a problem described in everyday terms into a form that the computer can accept and process, and the vast difference between these two forms is what lends the computer its air of mystery.

On the other hand, the work of transferring between the world outside the computer and the world within is one of the attractions for the programmer. He has developed the ability to mediate between two kinds of thinking—the one imprecise, complex, ambiguous, filled with connotations and metaphors and the other coldly precise and formal, without ambiguity, and therefore capable of perfect clarity. There was once a great effort to enable the computer to translate between human languages (from Russian to English for technical treatises), but really the situation is just the reverse. The programmer is the computer's indispensable translator. First he translates the language of the outside world into computer language and then the computed results back into a language that the world can understand and use.

The work of translation is the most strenuous and creative that can occupy a programmer. Some problems transfer relatively easily from one world to the other (some mathematical work, for example), but many do not. The building of computer models for social and economic purposes, such as the study of traffic flow through a city or the economic planning for a state, is often extremely complicated. At first, the greater the complexity, the more stimulating the challenge: any programmer would rather work on a program to model pollution problems than on a payroll program for an electric company, regardless of his political convictions. But there comes a point at which the difficulties of programming outweigh the rewards, emotional or monetary. Many problems cannot now be meaningfully programmed because they will not allow themselves to be circumscribed by the rules of electronic thought. An outstanding example is the problem of producing and understanding sentences in fluent English or in any other natural language. By any reasonable definition of understanding or writing English, no program yet exists that can do either. Successful programming is a movement between two modes of thought, modes so different that we cannot yet determine whether the most compelling human problems can ever be expressed in the formal language of computers. Meanwhile, the problems that do admit of an electronic solution are varied and important enough to keep millions of programmers at their jobs.

The satisfaction that comes to these programmers is that of establishing a correspondence between some facet of the outside world and the computer. This is the same satisfaction that an applied mathematician feels when he sees that the bridge built to his specifications actually carries a load, or the physicist, when the particle he predicted actually appears in the accelerator. There is another joy that the programmer can experience, one sometimes stronger than that of correspondence. A problem to be solved, a program to be written, is for him a justification for entering the beautiful world of computer formalism. This world has attractions of its own, which can make the programmer reluctant ever to leave. Even after he has decided in general what sort of a program is needed (determined the correspondence), he must still cast the details into a computer language. Many do not see this second stage as a mere chore but rather as an opportunity to explore the internal coherence of the programmed solution, to see how the parts of the program can be fit together smoothly and in an aesthetically pleasing way. A programmer may polish his program just as a watchmaker polishes and ornaments his work for display. In this respect, the programmer resembles the pure mathematician, whose work is to explore the internal coherence of a particular formal system without bothering about the relation of this system to the world outside his study. Indeed, non-Euclidean geometry began as abstract play in the imagination of such men as Gauss and Riemann; it was not until the twentieth century that Einstein found a use for it in his general theory of relativity.

Computer specialists, themselves logicians or mathematicians, can also spend their whole careers exploring the properties of the formal programming languages, leaving it to others to find applications. The ordinary programmer, no great logician, still delights in making his program faster and more compact (writing in three lines of code what he had previously written in five); and he may spend more time polishing his solution than he initially spent solving the problem. Indeed, the hypnotic effect of programming can produce a kind of addict, in Joseph Weizenbaum's terms a "hacker," a compulsive programmer for whom the activity has completely overshadowed the problem: "Wherever computer centers have become established," he writes, "bright young men of disheveled appearance, often with sunken glowing eyes, can be seen sitting at computer consoles, their arms tensed and waiting to fire their fingers, already poised to strike, at the buttons and keys on which their attention seems to be riveted as a gambler's on the rolling dice. When not so transfixed, they often sit at tables strewn with computer printouts over which they pore like possessed students of a cabbalistic text" (*Computer Power and Human Reason*, 116).

The hacker spends hours at the console refining programs that have no real purpose or ones whose purposes are elusive, such as playing perfect chess or producing natural English. He is the alchemist or magus of the twentieth century, convinced of the importance of his new idea without being able to see clearly where it will lead. And so he can do no more than play solipsistically with the electronic building blocks the computer provides, hoping to tinker together something of great significance almost by random effort. Yet every programmer knows the fascination of playing the electronic game for its own sake, of searching for a solution that is both correct and elegant. The hacker caricatures a real programming virtue, that of making one's program clean and coherent.

Electronic Limits

I have just described the twin qualities of programming, correspondence and internal coherence; the former is the problem of entering the stylized world of computer thought, the latter that of functioning effectively within that world. The description has been a utopian one. I should now explain in more detail the limitations under which the programmer must operate in both

framing his problem and programming his solution. The ultimate
constraints are imposed by the logical, indeed philosophical, character of the digital computer. A Turing machine may only consider problems whose solution can be reduced to a step-by-step procedure; the number of steps must be finite. The problem must be converted to one of structure, in which binary symbols can be manipulated to produce the desired output. These rules of the game are usually not felt as limitations. They are the very essence of the computer, the source of its power. In the same way, a mathematician does not feel constrained by the fact that he must obey rules of logical inference in proving a theorem, because without the rules of inference, there would be no theorems.

There is another set of constraints that trouble the utopian world of programming. These constraints come from a source previously mentioned. In nearly every program, limitations of time, space, or both must be overcome. The reader may still be skeptical that any reasonable program could suffer from a lack of either, if run on a large, modern machine.

Before resolving this doubt, I should magnify the paradox. If a man drops a coin from a height a bit above his waist, it will take about half a second to reach the ground. In that time even a respectable minicomputer could perform five hundred thousand or more additions, whereas the time required for the fastest machines to fetch a word from memory is now measured in billionths of a second, and a new generation of superconducting computers may reduce that time by a factor of 100. As for space, an academic computer center may have millions of bytes in core memory, billions in disk storage, and trillions on magnetic tape. Anyone who reads science articles in the weekly magazines is familiar with the kind of numerical rhetoric I have been using and has long become impervious to such figures. The rhetoric surely leaves the impression that storage and time are two commodities the computer specialist seldom needs to worry about.

In fact, programmers direct most of their energies to conserving the time and space their programs require. To begin with, the problem itself may exceed the reasonable capacity of any modern computer, and this happens not only with exotic problems of only theoretical interest but also with ones engineers face every day. The simple program described above for sorting a list alphabetically consumes too much time to be used on large files. If an electric company wanted to sort the names of a million customers, the number of computer operations required might run into the trillions (millions of millions) and take a modest machine (at one million operations per second) days of continuous operation. Other, faster sorting methods have had to be developed.

The situation is worse still for a problem that a mathematically minded reader may recall from high school algebra—computing with determinants. An engineer may well want to solve a system of equations that has a determinant of twenty columns and twenty rows. The straightforward method learned in high school is easily programmed, but, even with a judicious arrangement of terms, the algorithm would require on the order of 10^{18} separate multiplications and might take the fastest current computer 400 million years of continuous effort (Pennington, *Introductory Computer Methods*, 281–82). In this case, there is another method that would reduce the time needed to a fraction of a second. The discovery of efficient algorithms is one of the tasks of numerical analysis, whose work was discussed in an earlier chapter. This work is motivated not merely by the aesthetic pleasure of solving a problem in the neatest way possible but more by the need to find any realistic solution.

In recent years, computer programs have become skillful chess players. They cannot yet beat a master in tournament play but can compete effectively against all but the best human players. The programs do not work as the layman might think: they do not methodically consider all possible moves and choose the best. No computer has time or space for such a strategy, and no computer ever will. David Levy, a master interested in the possibilities and limitations of computer chess, has written: "If the number of feasible chess games were not so enormous, a computer would be able to play perfect chess. It could analyze the initial position out to mate or to a mandatory drawn position at the termination of every line of look-ahead analysis. But the number of possible games (more than 10^{120}) far exceeds the number of atoms in the universe and the time taken to calculate just one move in the perfect game would be measured in millions of years" (*1975 US Computer Chess Championship*, 2). The amount of memory required might be equally staggering. No foreseeable increases in speed will put perfect chess within the temporal reach of the von Neumann computer. Current programs only search out and consider a tiny fraction of the possible moves, and they often overlook good lines for just this reason.

This last example, like the determinant problem, illustrates that nemesis of programmers, the threat of a "combinatorial explosion." In many problems, the amount of time or space needed threatens to increase exponentially or even faster with the number of elements considered. These rates of growth are even worse than Malthus's grisly population curve. The trouble stems from the systematic approach that the program follows and the fact that each new element of data interacts with all or most of the previous elements, thus multiplying the program's complexity. Exponential or other intolerable rates of growth may crop up in any application of the computer. There is a huge body of literature on the problem of program complexity—the time and space a particular problem will require, both under the best theoretical conditions and under the conditions of a working computer system.

The remarkable Babbage broached the issue a century ago: "Now it is obvious that no *finite* machine can include infinity. It is also certain that no question *necessarily* involving infinity can ever be converted into any other in which the idea of infinity under some shape or other does not enter." Yet Babbage hoped to overcome the limitation in practice. "It is impossible to construct machinery occupying unlimited space; but it is possible to construct finite machinery, and to use it through unlimited time. It is this substitution of the *infinity of time* for the *infinity of space* which I have made use of, to limit the size of the engine and yet to retain its unlimited power" (Morrison and Morrison, *Charles Babbage and His Calculating Engines*, 60). Babbage never built his Analytical Engine and set it to work. If he had, he would soon have realized that time is as precious as space and that engineers of his day would have no difficulty setting problems whose answers would require interminable centuries of cranking for his Analytical Engine.

The problem is inherent in the idea of a logic machine; it will not disappear in a few decades with improved technology. No improvement in the von Neumann design as we know it will obviate the limits of electronic digital processing. As one specialist has put it: for a computer "actually to count to 2^{1000}—even operating at the wave frequency of hard cosmic rays—would take longer eons than even our most cosmological astronomers like to consider" (Minsky, *Computation: Finite and Infinite Machines*, 115). Yet solutions to many problems in logic and engineering would require far more than simple denumeration. Someday a new means of computing may well be found, a replacement for the von Neumann machine invented, which somehow avoids the limits of time and space (perhaps through the technique of parallel processing—using a large array of microprocessors instead of one CPU). That discovery will change the electronic world just as the steam engine changed the mechanical world. But in the foreseeable future, programmers will have to work within limits.

The whole of their present work comes down to the fact that the computer provides only finite resources with which to solve problems that are ever threatening to become infinite. Without the constraints of time and space, computer programming would be a wholly different art because the power of the computer would then be limited only by the formal theory of Turing machines. The idea is so heterodox that no one can be sure what the electronic world would be like. Could a chess program be written that follows out every possible game to completion, so that it would beat or draw a grand master every time? Would it be possible to prove important theorems of mathematics by the same exhaustive method? Would elegance and even careful work in programming make any difference any longer? Could all information gathered by mankind be stored in one vast data bank, just as the electric company now handles its billing and customer inquiries?

It is not at all certain that most programmers would want to live in such a seeming paradise. The constant concern for the limitations of electronic time and space is as much a part of their craft as the concern for the properties of concrete or steel is a part of structural engineering. In fact, programs are brought into collision with limited resouces not only by theoretical considerations but by the programmers themselves, in the very choice of problems and in their ambitious solutions, which forever call for larger and faster machines. Programmers fall victim to the same syndrome as engineers who deal with more tangible materials, the desire to work their materials and their own skill as designers to the limits of performance. We see the syndrome in the design of new military aircraft, with each new plane vastly and often unnecessarily more complex than the last, and so prone to endless new failures. Such engineering is the latest expression of the same mode of thought that led architects in the Middle Ages to push the vaults of their cathedrals ever higher, until they fell. But the results have by no means always been negative, for this same striving has brought about the real technological progress of the past two hundred years in Europe and North America.

As already mentioned, computer designers and programmers have every reason to believe in progress. Their machines have

become much more capacious and have been been evolving at a
remarkable rate, becoming perhaps a hundred times faster every ten years. Yet the growing aspirations of programmers have outstripped even this rate. In the days of the first academic computers, programmers were delighted to have a machine that could solve fairly complex differential equations in a few minutes or hours, using a working memory of a few hundred bytes. At present they routinely expect to solve in seconds problems that would require decades of human calculation; they manipulate data millions of bytes in length; and still we often read in scientific literature that a given problem had to be simplified in order to run realistically on current computers.

Speed and capacity cost money, and this is perhaps the most pedestrian intrusion of the real world into the electronic utopia. Faster components used in building both processors and memory units require more exotic materials or more careful fabrication. Programmers, like all other engineers, must work under a budget, and computer centers set their fees in proportion to the amount of time and space used, so that money becomes the measure of a programmer's resources. It is seldom possible to treat the two electronic commodities separately. The use of time is often inversely proportional to the use of space in the machine. The more memory you have available, the faster you can make your program; if space is dear, you will need to design a slower program. This trade-off between time and space is a classic issue in engineering, like the trade-off between strength and weight in structural engineering, and runs throughout the computer world from the design of transistors (the faster ones are sometimes larger) to the programming of data bases (faster rates of retrieval usually require more space). The difference is simply that computer specialists deal in more abstract commodities than their fellow engineers.

All engineers play the same game, although the rules are seldom as clearly defined as they are for the programmer. The rules of logical inference encompass like a starry sphere the whole world of the computer. The exploration of this ethereal region has been the work of logicians such as Turing, but the meaning of Turing's logical limitations for the future of real computers is unclear. On the other hand, every programmer is conscious of the difficulties of describing common problems in the rigorous language computers understand. It is in the act of translation that the game becomes challenging, and it remains challenging as the

programmer works to polish his translation, always with an eye
for the time he has available and the amount of memory within
which his program must operate.

Creating by Hand and by Machine

Let me set the craft of programming in its historical context: it is
an instance of the general craft of engineering that has grown up
since the eighteenth century. The kind of creative work that engineers perform (their materials, methods, and goals) is rather different from any other the world had previously seen. Engineering
is among Western Europe's most original contributions to the history of civilization, and programmers are special even among engineers. But that is not to say that the work of the programmer
cannot be meaningfully compared with other kinds of creativity
in past cultures. In every era there have been men and women
who have developed techniques for turning raw materials into the
useful or beautiful products of civilization, and they are all colleagues of the computer programmer.

The Greek craftsman exercised his special brand of creativity within what I have referred to as a manual technology. He
worked close to his materials with simple tools and skilled hands
rather than with imposing machines, in which much of the skill
of creation is built into the mechanism. The materials themselves
were natural—flax, wool, wood, clay; the sense of separation
from nature was not great. Greek craftsmen are not available for
questioning, but if we ask people who today practice the old
crafts as hobbies (spinning or weaving cloth by hand or making
pottery), we are told they find the work satisfying and peaceful
precisely because of the simple repetitive motion and the sense of
cooperating with nature rather than working against it. Part of
our modern prejudice in favor of such "natural" crafts is a revolt
against the dehumanizing aspects of the Industrial Revolution;
this prejudice would not have been shared by the Greeks. Still,
the Greeks genuinely enjoyed the skillful working of clay and
wood and wool: the craftsmen themselves are silent, but the surviving pottery and vase painting is evidence of their satisfaction
in their work. The ancient poets, both Greek and Latin, often
sang of the wonders of weaving, metalwork, and painting, magnifying the talent they saw in the shops and ascribing it to gods
like Hephaestus or Athena. The shield Hephaestus made for

Achilles captured in gold, tin, and enamel the entire Greek world at peace and at war. Athena's weaving was just as awesome.

Along with the love of handiwork came a contentment with the technical limitations set by nature. By later Western standards, the ancients were complacent about improving their technology. The Greek craftsman was content to work with a technology handed down from generation to generation and generally changing only by a process of refinement. Such "revolutions" as the introduction of iron or the double-beamed loom made their effects felt only gradually. Aristotle, more sensitive than Plato to the commonsense philosophy of his countrymen, believed that every art evolved until it reached its natural end or telos. From that point on, it could only degenerate. (He included Greek tragedy in the assessment.) There was, then, a strict limit to what could be accomplished in clay or stone, and there was no sense in trying to transcend that limit. Greek artists and craftsmen, however, were remarkable in their ability to turn this constraint into a virtue. The strong sense of line exhibited in their painting and architecture is a physical expression of their philosophy of art, where art means both technology and high art, for the ancients did not rigidly distinguish the two.

Ancient craftsmen were not really silent, for they had the philosophers as well as the poets to speak for them. It is striking that when Greek philosophers are addressing such lofty questions as the nature of being and creativity and creation on a cosmic scale, their thinking is influenced by the work of the lowly potter or carpenter. They saw the problem of creation (of the whole world or anything in it) with the eye of a potter examining his lump of clay: the problem was to explain how order was imposed upon an original disarray, how one moved from chaos to cosmos. The answers to this question varied from the highly imaginative and quasi-religious suggestions of the pre-Socratic philosophers to the exquisitely detailed, logically sound, and totally wrongheaded theories of Aristotle. Common to all was a passion for order and balance.

Let me cite only the two most famous ancient philosophers. In his myth of cosmic creation, the *Timaeus*, Plato explicitly compared the work of his creator deity to that of a craftsman, a metaphysical potter or carpenter who forms the universe as a living, finite, spherical creature. Plato emphasized that the deity wanted to follow an ideal pattern in his creation but was prevented by the nature of the materials at hand. He could not make a perfect world, just as a potter could not make a perfect vase, for the clay would crack or the paint would fade. The fundamental problem faced throughout history by craftsmen and engineers, indeed, by all those who try to bend nature to their own purpose, is succinctly put in the *Timaeus*: "The world came about as a combination of reason and necessity" (48A, my translation). Reason made it as perfect as necessity would allow. It is ironic that the basic issue of all technology should have been so well explained by a philosopher who held technology in low esteem. Plato was especially sensitive to the problem, although it is possible that he had never stepped into a potter's workshop in his life. He spent much of his philosophical career wrestling with the intractable nature of the physical world, so disturbed by the world's imperfections that he was led to create a parallel, perfect world of ideas. Since Plato himself was not a craftsman, he apparently saw only the negative side—that the craftsman's ideal conception of a piece is sullied by the imperfections of the wood or clay he must use. The craftsman himself no doubt felt equally strongly about the virtues of clay or wood, felt liberated as well as restricted by his materials.

In any case, the idea of imposing patterns upon shapeless materials, influenced by the working methods of potters and carpenters, was central to both Plato's and Aristotle's thinking and therefore to the thinking of the ancient world. Aristotle went on to explain the changing world of experience as intelligible form imposed upon neutral matter; behind his philosophical jargon is a clear analogy to the way a potter turns shapeless clay into a vase. This form-and-matter analysis survived not only Aristotle but the ancient world itself, serving as a foundation for medieval philosophy. In general, Aristotle's cosmos with its hierarchy of concentric spheres—spheres for the sun, moon, planets, and stars, all carefully delineated in their size and motion and circumscribed by the divinity—is a philosophical tribute to the sense of precision and line of the Greek arts and crafts.

The Middle Ages saw many technical innovations such as crop rotation, improved ploughs and harnesses, windmills, waterwheels, and the clock. Still, many of the ancient crafts remained, and the new technologies needed centuries to impress a mechanical-dynamic way of thinking upon society at large. The medieval view of craftsmanship may not have been much different from the ancient; certainly the collective skill required by a Gothic cathedral, with all its sculpture and minor arts, was no less impressive than that required by a Doric temple. But Christianity did bring about a new view of creation on the cosmic level because of the omnipotence of the Christian God. God preceded his creation and remained separate from it. There were no pre-existing materials from which he was required to work; he created ex nihilo and knew none of the limitations of earthly craftsmen. The imperfections of the world could not be ascribed to a god who was trying for perfection but could not pull it off. God's resources were infinite. Gone was Plato's notion of necessity imposing upon reason. The Christian God was not much of a craftsman, and human craftsmen, who aspired to the divine by pushing the vaults of their cathedrals higher and higher, were reminded of their own limitations when the vaults collapsed, as they did on more than one occasion.

In Christian theology, then, divine creation was wholly different from human creation. God's absolute brand of creativity was so remote that in the course of centuries more emphasis came to be placed on the worldly creative activities that were within the human grasp. Eventually men began to think of God in their own technological terms, as a divine clockmaker who had tinkered together a clockwork universe. Soon after, God's place as a creator was usurped entirely by men, by the engineers of the eighteenth and nineteenth centuries. That is, with the elimination of God from cosmology, men ascribed the arrangement of the universe to natural forces. Their interest in creation then returned to the human scale: what the man of technology, the engineer, could achieve within nature. Lewis Mumford aptly remarks of the great mechanist Descartes that, "by turning man into 'a machine in the hands of God,' he tacitly turned into gods those who were capable of designing and making machines" (*Myth of the Machine*, 2:84).

The forerunners of these engineers are to be found as early as the Middle Ages, when the two cardinal qualities of Western technology, the use of clockwork mechanisms and the pursuit of power, were already emerging. The more sober forerunners built the verge and foliot clocks and refined water mills and windmills. The more imaginative, if less productive, were the alchemists, who combined elements from Jewish culture (the cabala), Greek culture (the Hermetica), and folk traditions of magic. What they had in common with the designers of clocks and mills was a desire to use technical know-how to harness natural power and indeed to overcome the limitations that nature seemed to impose—to create the homunculus, to conquer mortality with the philosopher's stone, to turn lead into gold. In another sense, they were anti-engineers, so disappointed with the natural world that they sought magic ways of making the world immediately perfect.

The engineers of the last three hundred years have struck a middle course between the earlier craftsman's acceptance of the limitations of his materials and the alchemist's rejection. Their mixture of mathematical science and commonsense has been remarkably successful. Unlike the Christian God, the engineer cannot create ex nihilo. He must use metals that are laboriously dug from the earth, but he can invent techniques to modify these metals, making them stronger and less brittle; for example, he can melt iron ore to separate the slag and then add controlled amounts of carbon to form steel. He need not be satisfied with the relatively weak or irregular sources of power provided by the wind and by running streams. Instead, he can use his scientific knowledge of vacuums to create the "atmospheric engines" of Watt and Trevitheck. Since he lacks God's advantage of being able to form his materials wholly to suit his logic or imagination, his task is to make the less-than-perfect materials at hand suit his needs.

Reason and Necessity

Plato's characterization of creation as a marriage of reason and necessity applies, I think, to all human technology. It seems to capture perfectly the nature of computer programming, perhaps because programming is itself a paradigm of technology, a combination of European and ancient ideas of making and creating, crafting and engineering. Like the ancient craftsman or Plato's deity, the computer specialist works with materials whose fundamental limitations he must accept. His universe is finite, limited in time and space. The importance of logical, clear design and the lack of unnecessary ornament are common to the Greek potter and to a good programmer.

On the other hand, programming resembles the Western European idea of creation with its radical separation of creator and created thing. If the programmer enjoys working alone, in correspondence with the machine through a terminal, he is nevertheless aware that the world of the computer is one that he himself can never enter. It is a world of invisibly small electrons moving

at enormous speeds. The programmer never sees his program
run—at most he is treated to the flashing of several rows of lights at a console. The time scale at which individual instructions are executed is so much finer than that of human experience that the programmer can have no subjective understanding of this aspect of his work. The only indication he receives of what has happened in the seconds or fractions of a second during which his program runs may be a printed sheet with some unenlightening letters and numbers. One of the greatest problems in perfecting a program is simply to discover what is really happening inside the machine. The programmer (and even the designer of computers) is farther removed from his materials than anyone before in the history of technology.

In general, the technologists of the twentieth century share the goals of their predecessors: to make the world of nature serve the material needs and desires of mankind. Most recently, we see indications that engineering is entering a new phase because of the burden of overpopulation and the waste of a consumer society. The engineer is now confronted with the problem of conserving scarce resources rather than endlessly expanding their use. He is given the task of designing alternate sources of electricity (solar, fuel cell, nuclear fusion) rather than designing ever larger coal plants. He seeks to replace scarce and expensive resources with plentiful ones. (One of the advantages of optical over electric circuits, for example, is that glass is relatively cheap, whereas the demand for copper is making wire expensive.) There is nothing new in principle about engineering the replacement or conservation of scarce resources. It is often said that a major reason for the increase of coal burning in England (and therefore coal mining and the use of the steam engine) was the progressive deforestation of the countryside in the seventeenth century. Charcoal became dear, so dirty, hard-to-mine coal began to look more attractive as a source of heat.

Still, just as engineering in general has never before occupied so important a place in society, so the engineering of scarcity and replacement has never been called for on so grand a scale. We seem to be running out of everything—from fossil fuels to metals to arable land and potable water. This must tell in the long run, especially because population everywhere but in a few countries in Europe continues to increase, putting ever greater demands on diminishing resources. The idea of the engineer as a man who makes much of little, who cleverly manipulates limited resources, can only become more compelling. It enters into the popular conscience with such slogans as "spaceship earth" and "small is beautiful" and indeed with the ecological movement in general. There is the growing realization that every human resource, technological or political, will eventually be called upon to help.

In this new creativity of conservation, the computer plays a key role, particularly through its flexibility, its capacity to simulate other machines and human or natural systems. A computer programmer works with nothing less than time and space themselves, albeit computer time and electronic space, which are peculiar forms belonging to our century and culture. The structural engineer designing a bridge must consider the properties of the steel and concrete he uses because the bridge itself is a configuration of arches and cables. Computer space is a far more general entity than any metal or polymer, and for this reason programs can be written to simulate almost any structural and material problem engineers encounter. Computer time simulates the passage of years; computer memory stores the information of the traffic and pollution problems of an entire city. In simulation programs, as in every program he writes, the computer specialist recapitulates in its purest form the problem of every engineer in a society of limited resources. He knows exactly what time will cost him and how much space he can use. He is constantly reminded of these limits, for they are the essence of his craft.

Electronic Play

It would be wrong to end a discussion of programming on a note of impending disaster. The disasters are indeed impending, but the very isolation of programming allows some relief. The lone work offers rewards that are in some ways removed from the purpose of the assignment; even the dullest or most serious program gives us opportunities to experiment with the machine's logic. The computer encourages a kind of playful trial and error, a manipulation of electronic possibilities, so that it becomes almost irresistible to view programming as the ultimate sort of game. It is after all Turing's game, governed by the rules of finite automata and the limitations of the electronic components. The programmer makes up further rules as he works, rules that define the permissible data structures and the manipulation of those structures. Each program is a game-within-a-game. Like a player moving pawns on a chess board, the programmer maintains absolute, and therefore almost disinterested, control over his electronic resources. "The computer programmer . . . is a creator of universes for which he alone is the lawgiver. So, of course, is the designer of any game. . . . [Computer programs] compliantly obey their laws and vividly exhibit their obedient behavior. No playwright, no stage director, no emperor, however powerful, has ever exercised such absolute authority to arrange a stage or a field of battle and to command such unswervingly dutiful actors or troops" (Weizenbaum, *Computer Power and Human Reason*, 115).

What kind of creativity is exhibited in a well-played game? It is only recently perhaps that we have been willing to think of gamesmanship as a creative act. Although we may admit the educational value of play for children, game playing still falls somewhere between serious technical creativity (making machines to do work) and high art, between work and utter leisure. We refresh ourselves with games in order to go back to work, perhaps rightly. Yet the computer programmer works by playing games. His kind of creativity is more limited than we expect from high art, for the rules of the game restrict the results too severely. Obviously the programmer is not free to make anything when he sits down at his terminal; he is going to make a program—to solve some mathematical problem, to store, retrieve, or process some symbolic information. Along the way, the rules of his game are both his ally and his enemy: his achievements are circumscribed by them, but without them he achieves nothing.

For the programmer, the heroic struggle against nature, which has characterized Western technology at least since the Middle Ages, dissolves itself into a benign contest in which the adversary is not so much the physical world itself but the almost metaphysical limitations of the electronic universe. That universe is partly natural (electrons in the end) and partly artificial (the science of symbolic logic). The programmer confronts the dichotomy between reason and necessity, yet he takes a more stoic view of the limitations that necessity imposes. Perhaps the creation myth yet to be written, in which the creator-deity is a grand programmer, will differ in important ways from the Greek and Christian myths, as well as from those of the Enlightenment and Marxism. The programmer-god makes the world not once and for all but many times over again, rearranging its elements to suit each new program of creation. The universe proceeds like a program until it runs down or runs wild, and then the slate is wiped clean, and a new game is begun.

This description is not far from one of the currently plausible creation myths of modern physics. The universe began with a great explosion, and now all matter is rushing apart but at an ever diminishing speed. Some day matter may halt and begin to move back together, culminating in an annihilating collapse. And then perhaps a new universe will be born. Electronic technology, of course, had nothing to do with the proposal of this theory, which was based on astronomy and quantum physics, but the ideas of creativity fostered by the computer may well help to determine the modern attitude toward this or other suggestions the physicists offer us. Past Western thinkers have usually been too serious to regard the world as a game in the mind of a playful god, although Eastern thinkers have often done so. Such a view would mean a fundamental change, an end to the idea of infinite progress and the infinite striving of the Western soul and a new model for the individual and his society.

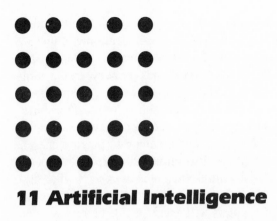

11 Artificial Intelligence

We return at last to the most radical expression of Turing's man, artificial intelligence—the notion of putting together hardware and programs to create new thinking entities, machines that rival human beings. In classical and Christian thinking, man was a made thing, the crown of creation, perhaps, but not the creator. Whatever modern biology has done to stress the continuity of life from microbe to man, we still think of ourselves as the highest manifestation of evolution, or the creative power of nature. In fact, our modern self-appraisal is possibly higher than that of the Platonist or the Christian theologian. They believed in orders of existence beyond our own, the Platonist in ideas and the Christian in angels and God. Dispensing with God as a hypothesis, the modern biologist sees man, and in particular his brain, as the most highly organized matter to be found in the natural world.

In his tiny artificial world, the computer programmer sets for himself the tasks of imitating nature and improving upon it, filling electronic space with models of real world problems and providing solutions through highly "unnatural" means. He invents complex transformations of numbers and symbols, and then has the satisfaction of seeing his results applied powerfully to the world of experience. Through mathematics, simulations, industrial robots, and data bases, more and more of human experience comes under the computer's "command and control." Is it surprising, then, that some programmers should want to rival the finest achievement of nature, to bring man himself into the computer world by turning the computer into an electronic brain?

The electronic brain remains for many an uneasy metaphor. No one can say with certainty how far the analogy between the computer and the brain may be taken—whether some human capacities, perhaps the most important, can ever be given to a machine. The popular press often carries reports of the computer's capacity for rational thought (from economic planning to playing chess), of its huge, infallible memory, its unimaginable speed of operation. Can computers really think? Computer experts, like laymen, are divided on this question. There are those who argue for various reasons that computers will never be able to think as men do, yet so little is known about how the human brain functions that their arguments are as speculative as Turing's original plea for machine intelligence. The argument over artificial intelligence has in fact produced more heat than light. The debate between two camps with such opposing world views leads quickly to a stalemate. One side claims that a computer will never be able to do this or that; the other replies that it will in two, five, or fifty years. Those who believe in artificial intelligence constantly exploit the fascinating, if frightening, uncertainty of our technological future. Who would be so foolish as to predict the limits of technology fifty years from now, assuming that science continues to make discoveries at its present rate?

For our purposes it does not matter whether computers can really think. We are interested in the cultural impact of the computer; for us the importance of the artificial intelligence movement is that it serves to crystallize so many of the qualities of electronic technology and display them in a way that will catch the imagination of our contemporaries. The debate over the possibility of computer thought will never be won or lost; it will simply cease to be of interest, like the previous debate over man as a clockwork mechanism. Computers will prove useful in many tasks and useless in others. It seems to me that the whole debate has turned the question around: the issue is not whether the computer can be made to think like a human, but whether humans can and will take on the qualities of digital computers. For that, as stated at the outset, is the fundamental promise and threat of the computer age, the fundamental premise of Turing's man.

Let us look more closely at the claim, made in 1950, that by the year 2000 computing machines would be capable of imitating human intelligence perfectly. Turing envisioned a game in which a human player is seated at a teletype console, by which he can communicate with a teletype in another room. Controlling this second console would be either another human or a digital computer. The player could ask any questions he wished through his console in order to determine whether he was in contact with a man or a machine.

Suppose there were in fact a computer at the other console. If asked to write a sonnet, the machine could attempt one or refuse; after all, most humans are not poets. If given two numbers to add, the machine might wait thirty seconds and provide the answer or instead might prefer to make a mistake to imitate human fallibility. However, it would not produce the answer in less than a second, for that would be a clear indication of its electronic nature. Turing's game really demands a machine that is more than human, not merely equal to its biological counterpart, one capable of any intellectual feat a man or woman can perform and sly enough to mask any prowess that exceeds a human's abilities. It would be a machine that knew men and women better than they know themselves. Turing was optimistic about the prospect of this supercomputer: "I believe that in about fifty years' time it will be plausible to programme computers . . . to make them play the imitation game so well that an average interrogator will not have more than 70 per cent chance of making the right identification after five minutes of questioning" (Feigenbaum and Feldman, *Computers and Thought*, 19).

The appeal of Turing's test is easy to understand. It offers an operational definition of intelligence quite in the spirit of behavioral psychology in the postwar era. A programmer can measure success by statistics—the number of human subjects fooled by his machine. The test seems to require no subjective judgment; it says nothing about the machine writing a good poem or solving an important mathematical theorem. Every humanist, of course, is tempted to devise his own Turing test and so his own definition of humanity: a computer will never be fully human unless it can laugh, cry, feel sympathy, feel pain, and so on. Someone has suggested that a computer will pass for a human only when it begins to ask what are the differences between itself and a human being.

Turing's own test is supposed to embrace any and all human qualities that can be communicated in writing. The player at the terminal may ask anything.

The test is cast in the form of a game, a duel of wits between man and machine. Games are in fact the form of intellectual activity that computers imitate most effectively. The Turing machine itself is a logical game, whose moves are governed by precise rules, and the computer plays a sort of game with every program it runs. Today, thirty years after Turing's proposal, a computer can play excellent chess, but no computer program could even attempt to play Turing's intelligence game. No computer could answer more than a question or two without revealing its mechanical nature.

The strategies for meeting Turing's proposal have varied. The most intriguing, if least successful, arose from the work of Norbert Wiener, who in the 1940s devised the term "cybernetics" for the "entire field of control and communication theory, whether in the machine or in the animal" (*Cybernetics*, 19). Wiener's work with servomechanisms to aim antiaircraft guns and to do much else besides had convinced him that forms of life could be understood entirely in mechanical terms; they could not be understood as Cartesian clockwork, which was too crude and rigid, but rather as electromechanical or even electronic devices. Like others, Wiener compared the new electronic tubes to neurons and wanted to subsume the study of both under one discipline. Wiener's outlook was clearly as much influenced by pre-electronic control devices (feedback loops in various machines) as by the digital computers just being built. In *Cybernetics* he stressed direct contact with the world—experiments with the muscles of the cat, improved prostheses for amputees, sensing equipment, and so on. Current workers in artificial intelligence show less interest in such direct contact with the world and more interest in abstract thought.

Wiener was still only halfway along the line from Descartes to Turing. He wanted machines to imitate the man who acts in the world as well as the man who reasons, to explain muscle action in terms of feedback loops as well as chess in terms of a digital program. He relied on hardware devices for his metaphor of man and demanded a close correspondence between man and the machine made to imitate him. Vacuum tubes were meant to be a physical substitute for neurons, servomechanisms for nerves acting upon muscles. This line of thinking was forthright and compelling, and led to attempts to build a brain (in theory, seldom in practice) using simple electronic components. Those following Wiener's approach spoke of creating artificial brain cells and neural networks and allowing the machine to learn as a baby was presumed to do—presuming with Locke that the baby's mind was a tabula rasa at birth. But the theory of neural networks, which was developed mathematically, met with little or no practical success. In general, Wiener's preferences gave way to others in the 1950s, as computer hardware and especially programming languages became more sophisticated. Unfortunately, the elegant name of cybernetics, created from the Greek word for governor but smacking perhaps of the antiquated technology of the war years, also gave way to "artificial intelligence."

Specialists more or less gave up the idea of building a machine whose components would mirror the elements of the human brain; they no longer demanded a literal correspondence between man and machine. The new high-level languages led them to emphasize programs rather than hardware, and they turned to such tasks as computer chess and theorem proving, problems of "information processing," rather than Wiener's command and control. In fact, the Turing test is just such a problem; it requires the computer not to act in the world but to act a role by manipulating symbols on a teletype.

For some, direct simulation of human thought seemed the most appealing way to pass the Turing test. They sought to discover intuitively how humans solved mental problems and then to translate these intuitions into digital programs. They may also have expected that the human solution would be the most appropriate (most efficient) one for the computer. Others tried simply to make programs fast and effective, feeling no need to be faithful to some theory of human cognition. Marvin Minsky, a principal spokesman for this approach, defined artificial intelligence as "the science of making machines do things that would require intelligence if done by men" (*Semantic Information Processing*, ed. Marvin Minsky, v).

This new definition seemed to reassert the difference between men and computers. Men can solve problems in one way, machines in another. But in fact, the analogy remains firm in the minds of programmers. Computer programs are open to inspection, and human ways of thinking are not. When a programmer devises an algorithm for playing chess or for analyzing English grammar, he can hardly avoid regarding human performance by analogy with his visible, intelligible algorithm. As one psychologist has put it, the computer model of the mind is the only working model available and even a bad model is better than none.

The nature of artificial intelligence can be illustrated by the performance of a program called SHRDLU. It is not the most recent effort (now more than ten years old) but one of the most famous and suggestive. SHRDLU is a simulated robot. The human controller types commands in simple English at a terminal; the program responds in typed English while a television screen displays the actions of the robot, picturing for us the tiny world in which the machine is supposed to operate. The screen shows a group of building blocks with identifying colors and a hook by which the robot moves the blocks. These elements constitute the program's world—all that SHRDLU can talk about or manipulate. The level of abstraction, the removal from the world, is striking. Nothing is really done by the robot, no blocks really moved; rather, the computer creates for us a picture of what it would do. A sample of the dialogue is given in figure 11-1. The human operator's questions or commands appear in small letters, and SHRDLU's replies appear in capitals. The program takes from five to twenty seconds to reply and to alter the display screen to reflect the commands.

SHRDLU can also remember when it has moved a block and why. It has a vocabulary limited to a few hundred words relevant to its world of blocks, and it can learn a few new definitions. If told that a "steeple" is a stack of two green cubes and a pyramid, it can construct one. It can parse sentences well enough to interpret commands and answer questions about its display of blocks in simple phrases. The ingenious programming behind these achievements, however, cannot be extended very far beyond the world of blocks. The program depends upon a complex data structure to represent blocks and their manipulation. Expanding that structure would make the program too large and too time-consuming for any contemporary machine.

It is hard to criticize the performance of this computer program, for the idea of any machine being able to perform seemingly intelligent tasks is at once paradoxical and compelling. SHRDLU was constructed in about 1970. Should we expect much more after twenty years of effort in artificial intelligence or be surprised at what SHRDLU can already do? The program is

Figure 11-1. SHRDLU

195
Artificial
Intelligence

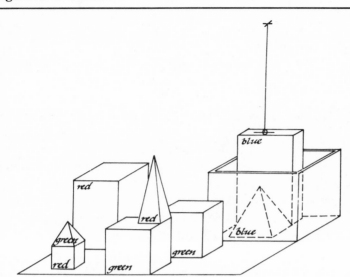

1. Pick up a big red block.
   OK. (The system answers
   "OK" when it carries out
   a command.)

2. Find a block which is taller
   than the one you are holding
   and put it into box.
   BY "IT", I ASSUME YOU
   MEAN THE BLOCK WHICH
   IS TALLER THAN THE
   ONE I AM HOLDING.
   OK.

3. What does the box contain?
   THE BLUE PYRAMID AND
   THE BLUE BLOCK.

4. How many blocks are not in
   the box?
   FOUR OF THEM.

5. Can a pyramid be sup-
   ported by a block?
   YES.

6. Can a pyramid support
   a pyramid?
   I DON'T KNOW.

7. Stack up two pyramids.
   I CAN'T. (The robot tries
   and fails.)

8. I own blocks which are not
   red, but I don't own anything
   which supports a pyramid.
   I UNDERSTAND.

9. Do I own a box?
   NO.

10. Do I own anything in the
    box?
    YES, TWO THINGS: THE
    BLUE BLOCK AND THE
    BLUE PYRAMID.

The world of toy blocks manipulated by the program SHRDLU. This sample has
been assembled from excerpts from Terry Winograd's description in *Understanding Natural Language* (New York Academic Press, 1972), 8–15. Comments in parentheses are Winograd's own. Sample reprinted by permission of
Terry Winograd and Academic Press.

certainly more an expression of faith in what artificial intelligence may someday achieve than a practical result in itself.

SHRDLU is a product of the intellectual assumptions of its creator. It is an invention, not a discovery, and it can be understood in the symbolic, psychological, and generic terms appropriate to any literary or artistic invention. From this point of view, SHRDLU is a retarded sibling of the robots known to us from decades of science fiction books and films. Robots in motion pictures evolved from such figures as the good Indian companion in westerns and the chimpanzee in the Tarzan series, figures who mediate between nature and full humanity or between savagery and full civilization. SHRDLU also mediates, in this case between the rigorously logical world of the computer and the ambiguous world of everyday human experience. It seems that we must meet SHRDLU more than halfway, giving up much of the richness of our experience and language in order to communicate with the machine. But in fact, SHRDLU has come a tremendous distance from its customary jargon of machine instructions even to achieve English that smacks of automation.

SHRDLU (indeed most programs that answer questions and solve problems) resembles Tonto or Tarzan's chimp in another respect: it exists to serve man, to help him manipulate his physical or intellectual environment, to carry out his requests patiently and supply him with information. Having no function or goal independent of the human operator, it embodies the stimulus-response psychology of the behaviorist, with the human always supplying the stimulus. Yet the structure of question and answer is the same Turing envisioned for his test. SHRDLU is already a game, although it is not Turing's game. Even if the range of responses were immensely broadened and SHRDLU could answer questions from any field of human knowledge in grammatical English (as it can now with its tiny world of blocks), no one would take SHRDLU for a human being. The naive submissiveness and total earnestness belong to a robot and not to a man or woman. Still, there is something disquieting about even this simple-minded machine. Entering into a dialogue with SHRDLU and agreeing to play the game, even under the restricted rules that the program understands, provides us with a moment's uncertainty and allows us to imagine that we are conversing with something that shares our humanity. This is the real importance of SHRDLU as well as the source of its appeal.

Probably the first step in meeting Turing's challenge is to create a program that is able to read and write English or some other natural language. SHRDLU is designed to demonstrate how a computer may "process natural language" in order to solve problems. One of the earliest aspirations of programmers was to bridge the enormous gap between human languages and the codes in which programs had to be written. As stated earlier in the chapter on language, progress was rapid in giving programmers more tractable coding languages, first assemblers and then compilers, but programs written in such languages are still far from simple English prose. Those interested in artificial intelligence dreamed of allowing their computers to accept unadulterated natural language. Their enthusiasm here is not surprising, for the electronic culture, accustomed to equating thought and language, could hardly regard a machine as intelligent unless it could speak a language fully as rich as English. Furthermore, a program that processed English would be able to communicate with those outside the scientific and engineering communities on their own terms; no one could excuse himself from confronting the thinking computer by claiming ignorance of the secret codes with which the machine functioned. To legitimate itself as an artificial intelligence, the computer simply had to learn English.

The first project along this line was the ambitious one of translation by machine: programming a computer to write an idiomatic English version of Russian prose. The military and scientific utility of such a program was obvious, so the United States government supported the work massively, spending perhaps twenty million dollars, until in the mid-1960s the task was judged hopeless. By then one of the specialists had the sobriety to write: "The outlook is grim for those who still cherish hopes for fully automatic high quality mechanical translation" (A. G. Oettinger, quoted in Dreyfus, *What Computers Can't Do*, 4). Those who did cherish such hopes changed their approach. They wrote programs to handle only a restricted number of sentence types or the highly restricted vocabulary of a single subject, such as the game of chess or a child's set of blocks. Their goal was no longer translation into another natural language but rather the "understanding" or "processing" of a single natural language.

Another area of continuing interest is that of human memory.

This is one of the most appealing human faculties to attempt to imitate. The computer is a device for storing and retrieving information, and human memory can be viewed in the same way. Many cognitive psychologists have been intrigued by this comparison, and the result has been a variety of programs for "simulating" memory. Such programs are complicated affairs; they are not of course designed simply to enable the computer to memorize symbols by rote. For that purpose, no special programming is needed because a magnetic tape or transistor memory in good repair preserves perfectly the information written on it; in this simple sense, it remembers with far greater accuracy than humans do. There is more to human memory than the ability to repeat what is remembered. If men and women are constantly forgetting what they learn, they can also remember more than they learn. They can trace out connections among sets of disparate memories and not only on the aesthetic level of Proust's associations on the scent of madeleines. Memory, with its capacity to establish structures of associations, is closely tied to other faculties of reasoned thought and creativity. It is in this sense that we live in the world we remember, and it is this mysterious capacity that psychologists and artificial intelligence specialists would like to co-opt for their computerized intellect.

The desire to reflect human memory in the circuits of a digital computer finds its most ironic expression in programs that forget as well as remember, ones deliberately designed to lose the address of some of the information fed to them and so to mirror human memory in its weakness as well as its strength. One of the favorite devices of psychologists of memory is some variant of the syllable test originated by Ebbinghaus in the nineteenth century. Human subjects are asked to memorize a list of syllables and are then tested in various ways—how many elements do they remember after five minutes or five days, can they reconstruct the order of the list, do they remember items at the end of the list better than those in the middle? From these questions, psychologists draw admittedly rather limited conclusions about how human memory works, about the size of short term memory, the difficulty of storing an element in long term memory, and so on. The whole procedure is an attempt to turn human subjects into quantifiable processors of information. Now programs for memory simulation reverse the procedure. The psychologist administers to the machine the same tests that have already been given to human subjects. If the program remembers and forgets in a pattern consistent with human subjects, then it is taken to be a fair model of human memory.

The effort to simulate the flaws of human memory is only half the story. Most artificial intelligence programmers are in the long run interested in perfecting the human reason, not imitating its imperfections. What they want to imitate is the human memory's capacity for association, its ability to retrieve appropriate memories instantly, apparently without recourse to the slow techniques known so far to the designers of data bases. They want to couple this capacity with the computer's perfect accuracy and the enormous size of its internal and external storage. The result would be an imposing resource upon which the logical calculus of the machine could draw. Many say that "representation is programming," that if we know how to represent and structure the information, then we have solved the problem. In this sense, memory might be the key to artificial intelligence.

Let me come back to the notion of problem solving. It is as important to the artificial intelligence specialist as to any other programmer. Artificial intelligence programs are not written to meditate; they are written to unravel some puzzle. Beyond the puzzles of the language and memory, artificial intelligence programs most often concern mathematical puzzles: manipulating formal symbols to prove theorems in logic, solving storybook algebra problems, or performing integration and differentiation analytically. Such programs have achieved some success by placing rigid restrictions on the puzzle questions they will accept. Again, those that permit a wide vocabulary of subjects remain extremely limited in the kinds of reasoning they can conduct. If you tell a typical program that the ball is in the box and that the box is in the room, it can then decide that the ball is in the room by the simplest form of set theory.

Computers have been successful at imitating humans when playing games, for games are generally conducted in a restricted universe of playing pieces and with an explicit set of rules. In the 1950s, a program to play checkers was able to match the talents of the best human players. The program used a set of mathematical linear equations to determine the desirability of moves. Because checkers is a relatively simple business, the program could "look ahead" for several moves, examining possible responses of its opponent and its own further responses. Success here encouraged artificial intelligence programmers to redouble their efforts with the more difficult game of chess. But the much greater complexity of chess, the combinatorial explosion of possible moves and responses, made the trick of looking ahead much more difficult to program. Chess programs face the fundamental limitations of computer time and space. Huge trees listing possible moves and replies have to be searched in *real time*, that is, while the opponent waits for a response. Improvement has been slow but steady since the 1960s, benefiting particularly from improved hardware such as faster CPUs and more capacious memories and from new techniques for storing and searching the data. Now, in the early 1980s, the best programs on the best machines can beat any talented amateur.

At present, artificial intelligence programs fall in these two groups: those that perform one clearly defined task at human or nearly human standards and those that perform more general tasks at drastically subhuman standards. In addition to chess programs, there are specialized algorithms to aid chemists in mass spectrography or doctors in diagnosis; these are impressive in their capacity to manipulate a large data base of expert and carefully defined information. On the other hand, SHRDLU deals with the tiny world of colored blocks, and already its wooden syntax, repetitiveness, and cheerful literal-mindedness make it apparent that we are not confronting an adult human intelligence. The program strikes us rather as an idiot savant or a well-behaved five-year-old child, for we very soon realize that, even in the simple manipulation of blocks, it is operating at the limit of its capacities. In fact, the world of artificial intelligence programs is populated by idiot savants and well-behaved children, but the goal remains to create machines that act with at least adult competence on the whole spectrum of problems requiring intelligence.

My own feeling is that the idiot savants will be with us for a long time, that computers will improve spectacularly in some areas and remain quite awkward in others. Computers may play brilliant chess in twenty years and still be incapable of translating English into idiomatic Russian. They might be able to process television images well enough to drive an automobile and yet not be able to prove important mathematical theorems automatically. And it is great fun, but probably fruitless, to try to guess which tasks will prove hard or perhaps impossible for the machine.

Successful or not, the artificial intelligence movement is important. So far its enthusiasm has gone far beyond the limits of its achievements and has led occasionally to wild claims, which later had to be retracted, to the delight of the polemicists against

artificial intelligence. There are practical benefits from work in
artificial intelligence: for example, intelligent robots to perform dangerous or unpleasant jobs in industry and to explore outer space. If in the future we can program computers in some simple form of English or use them to organize and search libraries of scientific and humanistic texts, we will likely be using techniques developed by programmers in artificial intelligence. Yet the literature of the movement often gives the impression that practical applications are of secondary importance, that a project ceases to be artificial intelligence as soon as it becomes practical. What really matters to the artificial intelligence specialist is to realize in transistors the image of a thinking human being: to make an electronic man.

The Technology of Making Man

There was perhaps never a moment in the ancient or modern history of Europe when no one was pursuing the idea of making a human being by other than the ordinary reproductive means. The pursuit today is more costly and demands a degree of mathematical sophistication as never before, but in fact the cultural equivalent of artificial intelligence can be found throughout the history of Western cultures.

The Pygmalion theme in Greek and Roman mythology shows how the ancients thought of going about it. Pygmalion was a master craftsman, so skilled as a sculptor that he could fashion the perfect likeness of a human woman. "He gave the ivory a form more beautiful than any mortal woman, and then fell in love with his own creation. The face was that of a real woman, who you would have thought was alive and—if her modesty did not prevent it—wishing to be touched" (Ovid, *Metamorphoses*, 10: 248–51, my translation). As Ovid, himself a classic craftsman-poet, described it, the statue, later called Galatea, resembled a human only in outward appearance; it was, after all, homogeneous ivory, with hair and paint applied to improve the deception. (In this painting, too, it resembled women of Ovid's day.) Pygmalion did not attempt to carve organs and blood vessels into his work, for ancient myths did not worry about such details. The subtle interrelation of Galatea's internal organs was not of interest, partly because a myth is not a medical textbook, but also because to the ancient mind (even in the sophisticated times following Hippocrates) the human body was not regarded as a complex of parts analogous to a clockwork mechanism. The four humors theory, though elaborated in later ages and particularly in the Renaissance, was first developed by Greek physicians. For them, man was a crucible in which the blend of black and yellow bile, blood, and phlegm determined the temperament as well as the health of the individual. If the ancients had a holistic view of man, the reason is that they understood very little of the complicated mechanisms and chemistry of respiration, circulation, elimination. Health and indeed life to their thinking was much more a question of the flowing and ebbing of vital liquids than of the careful regulation of biological processes.

So in the myth Pygmalion created only the form of a human; it was the goddess Aphrodite who breathed life into the work. The myth corresponded perfectly to the level of technology achieved by the Greeks and the Romans. The craftsman, the potter or the sculptor, concentrated his effort upon perfecting the form that he gave to his basic material (clay, stone, ivory). Pygmalion achieved such excellent results that he "fell in love with his own creation." His perfection of the human form simply demanded to become fully human, and the dea ex machina obliged. Ancient technology and a limited knowledge of biology inclined even philosophers to take an animistic view of the world. The deepest thinkers found it hard to reject the idea that anything that moved was alive; even their stars were made of a sort of living fire. Pygmalion's problem was that his craft was incomplete. He could not himself impart the final touch, the breath of life. Divine craftsmanship was needed.

In Plato's creation myth in the *Timaeus*, the deities who made men were clearly thought of as craftsmen (potters or perhaps bakers): they knew how to fashion the human vessels to contain the spark of life, and they knew how to add the spark to the mixture. In fact, the theme of the master craftsman, who had such perfect command over form that he could call forth motion and therefore life, emerges here and there throughout the literature of ancient mythology. The god of craftsmen, Hephaestus, could fashion gold and silver watchdogs and even female servants of gold. The mythic Daedalus could endow artificial wings with the power of flight. Neither the philosophers nor the poets had to provide details, to describe mechanisms. These feats were not conceived by the ancient mind in terms of machines but rather in terms of forms and animating spirits.

The Greeks of the Hellenistic Age did know of the possibilities of mechanical transformations of power. Mathematicians, among them Archimedes, studied the geometry of the five simple classes of "machines": the lever, the wedge, the screw, the pulley, and the winch. But the knowledge of mechanics seldom had a significant economic and social impact, for, whatever the mechanism, there was normally an animal or a slave at the business end. Complex mechanisms were generally conceived as toys. A surviving treatise by Hero of Alexandria describes such automata, various devices giving the illusion of animation. Some devices had the appeal of parlor tricks, such as a vessel that pours wine, water, or a mixture of the two. Some are ghost mechanisms—a temple whose doors open automatically when a fire is lit at the altar and close again when the fire goes out. But the most intriguing are the true mechanical representations of living creatures, men or gods. One was a theater in which the god Dionysus came forth, sprayed water and wine from his staff, and then was surrounded by Bacchants who danced in his honor. Like modern prime movers, Hero's mechanisms were powered by inanimate sources: falling water, heat, and atmospheric pressure. Hero even knew of steam power, which he used to drive a toy reaction jet known as an "aeliopile."

Thus, what later ages would regard as major sources of useful energy, the ancient used to power toys. The appeal of these toys was the paradox of motion without life; such motion had to be paradoxical to a society that was accustomed to thinking of life and motion as inseparable qualities. Hero's mechanisms themselves were never very complex. They can always be explained in a few paragraphs of prose. Yet they did not need complexity to fascinate an audience for whom the very idea of inanimate power was a delightful contradiction.

Both traditions of artificial life—the animistic represented by the Pygmalion myth and the mechanical represented by Hero's automata—also existed in Western Europe from the Middle Ages on. The first tradition was vastly elaborated by the alchemists and magicians of various sorts. Alchemy began, after all, in the ancient world, though we associate the search for the philosopher's stone and the attempt to turn base metals into precious ones particularly with the Middle Ages and the Renaissance. Another great project of the alchemists was the creation of an artificial man, the homunculus—a creature also to be found in the Jewish cabalistic tradition as the golem. The Renaissance alchemist Paracelsus gave a recipe for the creation of the homunculus; it was a carefully prepared mixture beginning with human semen, which makes clear that he had no mechanical robot in mind. The alchemist did not work with gears and mechanisms but rather with elements, spirits, distillates, sublimations, and spells. He belonged to the intellectual undercurrent of his times, whereas the better-known philosophers were exploring other ways and trying, not always successfully, to keep clear of ideas of animism and magic. Occasionally, particularly in the Renaissance, a magus like Ficino or Bruno emerged to make a lasting contribution to Western thought, but the belief in a world populated by spirits and subject to the laws of natural magic ran counter to the mainstream. The homunculus always had a certain eerie charm, even for poets like Goethe who knew better. But even in times when the majority of a superstitious population may have considered the homunculus possible, the idea exercised only a limited influence upon philosophy and literature.

Far more influential were the mechanical automata that began to be built along with the first clocks in the thirteenth century. The great Strasbourg clock, for example, had three Magi and a cock that crowed at dawn. Many have noted the influence of the clock upon the great mechanical philosophies of Descartes and Leibniz. It was not only the clock itself that sparked their thinking but also the thriving tradition of clockwork animals and men. The great civic clocks from the later Middle Ages were often decorated with moving figures, and the art of making automata, separate from or together with clocks, reached a high state of refinement in the centuries of the scientific and industrial revolutions. Descartes himself remarked that his mechanical explanation of animal life would make perfect sense to those who know "how many different *automata* or moving machines can be made by the industry of man" (*Philosophical Works*, 1:116).

Many such devices were designed to amuse royalty. In the sixteenth and seventeenth centuries, gardens were adorned with hydraulic automata, such as those described by the French engineer, Solomon de Caus. These Baroque toys, far more intricate than those of Hero, were unwitting tributes to the waterwheel (little used by the ancients) and the mechanical clock (probably unknown to the ancients). Ironically, ancient themes remained popular: grottos were constructed with such pastoral characters as nymphs, shepherds, and the Cyclops. About 1600 St. Germain in France could boast a nymph playing on a water organ, Mer-

cury sounding a trumpet, Neptune driving a chariot of sea-
horses, and Orpheus with his lyre charming the animals around
him—all powered by falling water.

By the eighteenth century, automata had dried out; that is, the
most interesting were spring-driven mechanisms. They had also
moved from the garden to the salon and could now copy words
and play music. Jacquet-Droz, a French toy maker, produced a
mechanical boy that would dip its pen in an inkwell and write a
message, distinguishing between light and heavy strokes and lift-
ing the pen between words and over the line. The mechanism inside could be adjusted to produce different messages. This same
toy maker built a female musician who actually played the harpsi-
chord with artificial fingers and at the same time breathed, raised
her eyes, and turned her head; she may well have performed at
the courts of Louis XV, Louis XVI, and George III of England.
One of the best known of the eighteenth-century automata was
the duck designed by Jacques de Vaucanson. Exhibited in 1738,
this creature drank, ate, quacked, splashed about, and even eliminated its food. It was also a sign of the progress of the Industrial
Revolution that the duck not only appeared before royalty but was
sold to exhibitors who took it throughout Europe.

Surely ancestors of Vaucanson's duck helped to confirm Descartes in the belief that even the most complicated physical processes of animals and men could be explained as intricate clockwork mechanisms. Only the mind was exempt from Descartes's
mechanical explanation, but in the eighteenth century a few men
were so impressed by technological progress—on the sober side
by the accurate watches and other precise instruments that were
being produced and on the lighter side by the toys of Jacquet-
Droz and Vaucanson—that they were prepared to go further and
see the whole man as a clockwork mechanism. Such was the the-
sis of the infamous philosophe La Mettrie, who wrote: "Let us
conclude bravely that man is a machine; and that there is in the
universe only one kind of substance subject to various modifica-
tions" (from *L'Homme-Machine*, cited in Vartanian, *La Mettrie's
L'Homme-Machine*, 197). La Mettrie was bold indeed, bold
enough to banish mental substance altogether in favor of the
material.

With the mechanical triumphs of the eighteenth century, the
situation was completely reversed from that of the classical
world. There the animists won the day and were convinced that
man was a material, like clay or stone, animated by the breath or

spark of life. The myth of Pygmalion or of Plato's craftsmen-
deities captured perfectly the prevailing world view. Hero and his
mechanical toys expressed the minority opinion, arguing paradoxically against animism. But in Western Europe, the animists,
the alchemists who sought to make men from a recipe, were in
the minority. The mechanical view triumphed: the bodies of animals and men were best approximated by clocks and best imitated by clockwork toys.

The triumph became more apparent with each passing decade,
as automata of the nineteenth and early twentieth centuries became progressively more precise and complex. Eventually, electric circuits and motors created even more lifelike machines than
clockwork could alone. Convincing if somewhat creaky electromechanical robots, such as the chess player of Torres Quevedo, began to appear and to hint at the electronic age to come.
The Spanish technologist Torres himself showed remarkable historical sense when he said in a 1915 interview, "The ancient au-
tomatons . . . imitate the appearance and movements of living beings, but this has not much practical interest, and what
is wanted is a class of apparatus which leaves out the merely
visible gestures of man and attempts to accomplish the results
which a living person obtains, thus replacing a man by a machine" (cited in Eames, *A Computer Perspective*, 67). Torres was
close to the definition of artificial intelligence, although his own
work was necessarily limited to mechanical rather than electronic
techniques.

The Electronic Image of Man

It is in the context of the classical and Western European traditions of making men that artificial intelligence must be understood. Artificial intelligence is not a science, any more than the
sculpture represented in the Pygmalion myth or the toy making of
Jacquet-Droz was a science. It is rather a special skill in engineering and logic; computer programming techniques are used to
mimic human abilities, from playing chess to answering questions in English. The techniques are different from those of the
mechanical men of the eighteenth century or of Greek myth, for
the simple reason that artificial intelligence is meant to take account of the new qualities that the computer has introduced into

the story of technology. That is its significance. Artificial intelligence is a radical expression of the possibilities of the digital computer, a celebration of a new technology.

Specialists in artificial intelligence must refer to their field as a science because we live in an age of science when practically every activity can only be dignified, or indeed legitimated, under this name. For this reason, they often claim their program is a "model" of this or that aspect of intelligence and point out that many sciences use models to understand nature. The problem here is that the artificial intelligence specialist has nothing but a model. Having abandoned the idea that electronic circuits can be made to mirror the organization of human neurons, he has no natural phenomenon left to study. A physicist may build a mathematical or even physical model, but he must at some point go back to nature for confirmation. But what can the computer programmer find in the brain or the mind of which his binary coded instructions are a model?

All the biologist can find are neurons, and the tangle of billions of axons and synapses will remain beyond human comprehension probably for decades. So the artificial intelligence specialist looks to "higher levels of the system," to problem-solving structures, rules of language production, symbolic representation of visual data, and the like. But beyond the level of neurons, there *is* no science of the mind; instead, there are metaphors that capture more or less aptly our mental experience. Programmers find in the mind structures appropriate to computers and formal logic because they bring these concepts with them as they look at human experience. Each age has found in the human mind precisely what it has brought to the search; each has had its own metaphorical explanations. Plato compared the mind to the class structure of a Greek city-state, the mechanists likened it to a collection of gears, and now Turing's man sees it as a digital computer. Certainly some metaphors are better than others. The mind seems more like a computer than like a clock. But some metaphors emphasize such disparate qualities that they can hardly be compared: is the mind more like a computer than a Greek city-state? The important point is that the computer is not the final, correct answer. It will surely be replaced someday by metaphors that spring from a technology we cannot now imagine.

So artificial intelligence will not be a science even if it wins someday at Turing's game. It is and will remain a field of engineering that relies on science and mathematics for its tools. Many of its exponents are computer specialists who have made real contributions to their discipline. (Vaucanson also made substantive improvements in the manufacture of silk, in addition to his automata.) Programs for artificial intelligence are not scientific constructions at all; they are wonderfully clever ornaments, like the cock and the Magi on the original Strasbourg clock.

We must not underestimate the value of such decorations, however, because the figures on the Strasbourg clock caught the contemporary imagination more readily than the concealed mechanism that actually told the time. In the ancient world, Hero's mechanisms had been mere toys because ancient technology was not mechanically inclined, but the writing boy of Jacquet-Droz and Vaucanson's duck were more than toys. They were leisurely statements of the idea that the natural world behaved mechanically; they made tangible the ideas of Descartes, Huygens, or Newton. The same ideas and the same creative energies also led to more practical applications, to better clocks and more efficient engines.

Similarly, the artificial intelligence movement produces programs that illustrate rather than perform. These programs are likely to have practical applications that even their creators may not foresee, but that is not their purpose. In a field generally marked by a ruthless sense of utility, these computer specialists are interested in something more marvelous than immediately useful results: they are seeking to demonstrate what it means to live in the computer age. Their programs succeed in arousing much the same wonder among the public, and even within the literary and scientific communities, that Vaucanson's duck aroused two hundred years ago. For example, the chess programs that play nearly at the level of the professional rate publicity whenever a chess master agrees to compete against a machine. Nowhere, perhaps, will people more readily grasp the meaning of electronic technology than in the widespread attention that artificial intelligence programs attract.

For the artificial intelligence programmer, the electronic brain is a dead metaphor: the computer and the brain differ only in the unimportant respect that one is made of electronic components and the other of biological ones. Both think. By taking the metaphor to its extreme, proponents of artificial intelligence illustrate with utmost clarity a way of thinking shared by all of Turing's

men. Other computer specialists or scientists who use computers
disagree sharply with the notion of artificial intelligence. Yet they all cheerfully speak of computer *languages*, the *logic* of computer circuits, and computer *memory*. They say that a computer did not *read* the data properly or *recognize* a particular character. They all accept to some degree the idea that humans and computers are comparable things, even as they assert that humans perform some tasks much better than machines. The comparison has become irresistible, a clear indication that we are living in the computer age.

The concept of artificial intelligence in fact contains within it most of the defining qualities of computer technology, qualities that have been explored in previous chapters. If a digital computer can think, then intelligent thought must be a step-by-step process conducted in pulses of time—data shuttled through a biological or electronic job shop in which each machine can handle only one bundle of data before or after another machine has had its turn. It must be conducted in a symbolic code, a "language of thought," directing the data through a series of transformations or intermediate forms between input and output. The transformations themselves must be governed by the rules of logic, which allow for the spatial and structural representation of meaning and intention. Such qualities of the computer shape every artificial intelligence program and every other program besides.

It is important to remember that these qualities are seen as strengths of the computer, not as limitations. The whole point is to find a way to enable a network of wires and transistors to manifest intelligence. Any artificial intelligence specialist could marry and produce a number of biological information processors in the customary way, as could a toy maker in the eighteenth century, a Renaissance alchemist, or an ancient sculptor. Turing, by the way, forbade this route; to meet his test, the machine must be electronic rather than biological. The attempt to make man over, with whatever available technology, is an attempt to circumvent or reverse the process of nature. Man the artificer and man the artifact merge, but on the artificer's terms. In bypassing the ordinary sexual process of reproduction, man achieves a new freedom from nature; computer technology offers a path to this new freedom.

The homunculus that results is man as a processor of information—not a whole man, for he has no arms or legs, nor emotions in any conventional sense. He is a calculating engine, although one of far more complexity and even charm than is portrayed in the popular mythology of sinister, electronic superbrains. He embodies assumptions about space, time, language, and creativity central to the computer era upon which we are entering. He is far more sophisticated than the stimulus-and-response circuit of the behaviorist school, but he does have the same relation to his environment, for he thinks by transforming inputs into outputs.

Most important, this computer man fulfills the same function for our contemporary technology that the clockwork man or the living statue fulfilled for previous ones. He is an expression of both the exciting possibilities and the limits suggested by this new technology. Some programmers in artificial intelligence are wildly optimistic about the possibilities. They speak as if at any time someone may find the key to making the computer not merely human, but superhuman. "Artificial intelligence is the next step in evolution," says one. And again: "I suspect there will be very little communication between machines and humans because unless the machines condescend to talk to us about something that interests us, we'll have no communication" (McCorduck, *Machines Who Think*, 346, 347). These artificial intelligences will help run society and relieve mankind of the burden of being the leading species. We have entered the realm of science fiction, and, as with all science fiction, the predictions are not really about the future but about an extrapolated present. Everything else is left constant or dropped from consideration while digital computers and their programs are allowed to grow arbitrarily in power. The world is seen through the electronic eye of the information processor.

The historian Lynn White has described for us a parallel from the Middle Ages. "By the middle of the thirteenth century," he writes, "a considerable group of active minds, stimulated not only by the technological successes of recent generations, but also led on by the will-o'-the-wisp of perpetual motion, were beginning to generalize the concept of mechanical power. . . . They were power-conscious to the point of fantasy. But without such fantasy, such soaring imagination, the power technology of the Western world would not have been developed" (*Medieval Technology and Social Change*, 133–34). Perhaps historians two hundred years from now (if history is still being written) will make the same assessment of the artificial intelligence movement.

Men tend to regard what they make through their technological or artistic genius as in some measure human. It is not only a skilled playwright who brings his characters to life: all human artifacts, from vases to computer programs, from tragedies to steam engines, have a bit of life in them. The artifact also changes the artificer. Through technology, men and women attempt to redefine their relationship to their environment on terms that seem more favorable. If the struggle to collect or catch food seems too difficult using only human hands and muscles, the addition of even a crude stone weapon or tool redefines the entire problem and opens new possibilities for personal and social activity at the same time. Human beings with a developed stone technology are different from those without tools; men with bronze and iron weapons, looms and potter's wheels, are different again. Each technology, if not each single invention, remakes the men who invent or possess it by altering their most elemental capacity— that of surviving in the world of nature.

If humans are not unique among animals in this respect, they are certainly very special. Most animals can rely only on the biological processes of evolution over thousands or millions of years in order to improve the equipment (teeth, camouflage, instinctual cunning) with which they must confront their environment. Animals that do learn to use tools and pass on such information to their young progress at an agonizingly slow rate, if they progress at all. In contrast, it is often said that humans progress too quickly to assimilate intelligently the technology they create. At any rate, the human race has taken over at least in part the future course of its own evolution by making itself over through technology.

So far only in part. Even today we are extremely limited in our ability to use chemistry or physics to control and improve our own human nature. We have at our disposal sources of energy on a cosmic scale: we can create explosions whose temperatures match those in the interior of the sun. Yet we still cannot make a prosthesis that has a fraction of the agility and versatility of the human arm. We know at least where to look for the mechanism that controls the growth and function of human tissue and organs—in the DNA in the nucleus of our cells. Yet these molecules are of such complexity that we can only decipher tiny portions, and our tools for rearranging molecules are correspondingly crude, although these very tools manifest the highest skill of contemporary biochemists. Men are clearly not in the position today to take full or even substantial control of their own biological destiny. And if this can be said of contemporary technology, it was surely the case with the technologies of eighteenth century Europe and Alexandria in the third century B.C. The machines of the Industrial Revolution performed only elementary repetitive actions; they had nothing comparable to the versatility of a skilled human worker. In fact, each of the cleverest automata of Vaucanson and Jacquet-Droz was only capable of a few mechanical tricks. The automata of Hero were far more crude, and in general the ancient world could only manage to use clay, stone, or bronze to produce static likenesses of human beings, although these likenesses were sometimes of extraordinary beauty.

Our ability technologically to improve upon our physical and mental equipment has fallen far short of the goal of remaking ourselves entirely. Nonetheless, technological man has always had that goal somewhere in the back of his mind, and it has emerged audaciously in myths, legends, toys, and automata throughout history. In all the manifestations, the informing idea is that the artificer and the artifact become one. Man makes man and therefore raises himself above the status that nature seems to have assigned him. Beyond that, each manifestation is different because it must, in Norbert Wiener's fine phrase, take account of "the living technique of the age." Ancient craftsmanship produced myths of ivory and gold statues coming to life and automata in which inanimate motion itself was a thing to be admired. Western European technology emphasized the intricate working of gears in its automata. The attempts at artificial intelligence today stress the qualities of contemporary technology that are new and strikingly productive.

There are antiquarians who still delight in the toys of the past; no one else today cares about making automata that physically resemble their makers, a fact Torres Quevedo recognized over sixty years ago. Engineers are indeed working to perfect robots (grasping and manipulating devices of various sorts controlled by microprocessors or full-scale computers). These devices have so many practical applications, for example, in handling dangerous materials or doing repetitive assembly work, that they have descended from the dreamy world of artificial intelligence into the world of business.

A so-called robot assembling machine tools bears hardly any

resemblance to the androids in science fiction and comic books of the preceding generation, the era before digital computers. One glance at these modern industrial robots convinces us that engineers are not seeking to make their machines look human, although they might have done so for aesthetic reasons. On the other hand, a programmer writing a mathematical "theorem prover" would say that he is interested in imitating a far more important and characteristically human quality, that of intelligence. One opponent of the movement has argued that the human body with its five senses is fundamental to the human capacity for intelligent thought. The argument misses the point. The artificial intelligence specialist is not interested in imitating the whole man. The very reason he regards intelligence (rational "problem solving") as fundamental is that such intelligence corresponds to the new and compelling qualities of electronic technology. Today, as before, technology determines what part of the man will be imitated.

12 Conclusion

Natural Man

The preceding chapters contain my sketch of Turing's man and explain which elements of computer technology form his peculiar character. It is now possible, I think, to address the questions posed at the beginning: Should we be repelled by the notion of man as computer? Is Turing's man a cultural aberration? It seems clear that he is the next step, the latest fashion, in our culture's image of itself. But is he a caricature of the great images of the past: Plato's ancient man, Christian man, enlightened man? Turing's man is the most complete integration of humanity and technology the world has ever seen. Have we gone too far in our attempt to make ourselves over through technology?

Critics of technology in general and of the computer in particular want to compare Turing's man to an ideal of humanity unaffected by the excesses of the machine. Man, they point out, is not a discrete information processor but a person who uses other faculties as well as pure ratiocination to give purpose to his life. He is in intimate contact with the world of nature through his senses, a contact that the computer cannot hope to achieve with its digitalized input and output of data. Lewis Mumford speaks for many when he writes: "Since the computer is limited to handling only so much experience as can be abstracted in symbolic or numerical form, it is incapable of dealing directly, as organisms must, with the steady influx of concrete, unprogrammable experience." This leads him to complain of the "utter absence of innate subjective potentialities in the computer" (*Myth of the Machine*, vol. 2, Graphic Section I, p. 6). Many others too have emphasized that men and women are capable of purposive action, of setting goals for themselves, and that this ability to form purposes assures mankind an autonomy that machines do not have.

This picture of natural man in contrast to electronic man is indeed an attractive one, for which I myself have great sympathy. But it is only fair to point out that such a natural man is hard to find in history. There never was a time when more than a small fraction of mankind in the West lived according to these standards of autonomy, love of nature, emotional balance, and the like. Critics of the new technology sometimes talk as if mankind always aspired to the ideal of being sensitive, nature loving, and creative until the computer was devised to rob it of these aspirations. In fact, this definition of the human condition is quite recent, the product of philosophy and psychology popular only in the last two hundred years. The values here assumed were not obvious to generations of ordinary people or philosophers.

Erotic feelings are part of the biology of the human race. Romance and sentimental attachments to nature, animals, and children are not. The Greeks, so often praised for their emotional balance, did not share our notions of romantic love, which first developed with the medieval custom of courtly love. Sentiment toward children was even slower to develop. Many Greek states sanctioned the practice of setting out unwanted offspring to die of neglect, and female children seem particularly to have suffered this fate. Even in Western Europe, children were usually depicted in art as tiny adults and thought of as incomplete human beings until the nineteenth century.

In the same way, the apotheosis of nature and man in nature is a product of the Romantic revolution. The Greeks had no such view. The great humanist Socrates, when asked by his friend Phaedrus why he seldom went outside the walls of Athens, said simply: "I am fond of learning. The country places and trees have nothing to teach me, but people in the city do" (*Phaedrus*, 230D, my translation). Learning from nature seems obvious to us; we complain that we get too few opportunities, forgetting that until recently men and women did all they could to get away from nature, huddling together in towns and villages for warmth and protection. Only the relative affluence of the last two hundred years and the appalling urban conditions brought on by the Industrial Revolution led Rousseau and the Romantics to want to reverse the trend. What of the restorative effects of feeling and emotion over against arid rationalism? It is worth recalling that Stoicism, perhaps the most popular philosophy in ancient times, preached the virtues of life according to universal reason, withdrawal from the world, and total freedom from emotion. Until at least the eighteenth century, Western European moral philosophy regarded the passions as something to be mastered, if not eradicated.

In other words, the ideal against which Turing's man is often measured is a recent invention. We may still prefer this definition of natural man to any other, but we cannot hope to freeze history: ideas and ideals of human nature will continue to change in the future as they have in the past. After all, according to this definition, all previous ages (including ancient Athens, France in the twelfth century, Renaissance Italy, Elizabethan England) were at least as dehumanizing as the computer age promises to be. Yet each previous age did have its ideals, its vision of the human condition; and these ideals were formed in part, like everything else in the culture, by the contemporary technology.

From Socrates to Faust to Turing

The ancient ideal was characterized by balance, proportion, a sense of sane limits in human affairs—an ideal announced humbly enough by the Greek potter (molding his wares with careful symmetry and decorating them in a spare, linear fashion) but reverberating throughout society and literature. The two famous Greek proverbs "nothing in excess" and "know yourself" are both admonitions for the life of balance and limits. Aristotle made the first the keystone of his ethical system, which set a mean between extremes in most human activities. The second, so closely associated with Socrates, was not a plea for psychoanalytic or Christian soul-searching: it meant instead that men must know their limits, particularly as mortals in relation to the divine, and be careful not to overstep them. What happened when men did overstep their limits was the favorite, indeed the definitive, subject of Greek tragedy.

With an appreciation of formal balance in art and life came the tendency to be superficial; and the Greeks would not have regarded this as a criticism. The idea of, in fact the obsession with, plumbing the depths of any experience belonged rather to West-

ern European culture. The ancients appreciated the linear, the superficial, the immediate, and the tactile in mathematics, art, and
ethics as well. Hence conscience played a relatively small role in
ancient ethics: crimes against individuals or the state and impiety
toward the gods were acts committed at a definite time and place;
sins thought but not acted upon did not matter. Another aspect of
the superficial was that the Greeks had comparatively little concern for history. The past did not weigh heavily upon them, as it
did upon the men of Western Europe, for the ancient world itself
had no ancient world against which to measure its achievements.
Nor did the ancient man care to look far into the future, for there
was nothing like the Christian concept of millennium or the secular notion of progress to direct his gaze. As Spengler put it, "The
Classical Life exhausted itself in the completeness of the moment" (*Decline of the West*, 1:267).

The limit of the ancient gaze was not only a temporal one.
Most philosophers had no difficulty finding the physical limits of
the universe: it was the sphere of the fixed stars, beyond which
there was absolutely nothing. In social terms, a contentment with
limited resources expressed itself in a steady-state economy and
in a disinterest in technological innovation. The inventor at least
had an honored, if not prominent, place in Greek mythology, but
the explorer was a character type seldom found in ancient literature or ancient history. Odysseus was not really an explorer, but a
wanderer, trying his best to get home, and Alexander the Great
marched from Macedonia to India as a conqueror, not an explorer. The Greeks were gifted sailors and brazen enough to face
the sea in small, wooden boats, yet they seldom ventured beyond
the Pillars of Heracles. How complacent they seem when judged
by the Western European standard.

In fact, the ancient character differed from the Western European in every respect I have mentioned. The Western European
character came, of course, in a dozen varieties, as did the ancient, but common to all was a fascination with depth, a desire to
penetrate the surface of reality. In art, this led to the invention of
linear perspective: the technique of organizing a whole picture
around a single deep focus, giving the illusion of three dimensions. In moral terms, it led to the Christian preoccupation with
the human soul as something deep and mysterious beneath the
facade of human behavior. In later, secular times, psychology
took up where Christianity left off, exploring the depths of conscious and unconscious human experience. This self-searching

was a principal theme of literature from the Middle Ages on: the quest for the Grail in medieval romance or for salvation in Dante later became the search for knowledge of oneself and the world in the nineteenth-century novel, as such critics as Northrop Frye have taught us. Nor was the drive limited to fiction. The great explorers of the fifteenth, sixteenth, and seventeenth centuries were answering the same call, as were the creators of modern science.

Spengler called this Western character "Faustian," after Goethe's rendering of the scientist-magician who seeks ultimate power and knowledge. He argued that the Faustian character was far more conscious of its place in history than the ancient man. And he emphasized, justly I think, the fact that Faustian men had an appreciation for the idea of infinity, which the ancients did not. The point is obvious in a comparison of ancient and modern mathematics, but it emerges in other ways, too. Because of Christian dogma, Western men were accustomed to regard infinity as a good thing (was not God infinitely good and powerful?), but the ancients often associated the infinite with something unintelligible and therefore evil.

Where does the computer age fit in this scheme? It constitutes another turning point, another major change in sensibilities. As I have argued, computer technology is a curious combination of ancient and Western European technical qualities. Developing through modern science and engineering, it nonetheless encourages its users to think in some ways in ancient technical terms. Turing's man has in fact inherited traits from both the ancient and Western European characters, and the very combination of these traits makes him different from either. Those of us who belong to the last generation of the Western European mentality, who still live by the rapidly fading ideals of the previous era, must reconcile ourselves to the fact that electronic man does not in all ways share our view of self and world.

In one fundamental sense, Turing's man has only taken Western European thinking one step further. He has forced to one extreme the dividing line between nature and the artificial. Throughout the industrializing period, our culture was busy making the world an ever more artificial place: making technologies out of matters that originally belonged to nature. Agriculture, metallurgy, textiles, and inanimate prime movers are all instances. With less and less that was unimproved, the result of a total artificial world, a complete change from nature to artifact, became thinkable. Samuel Butler stated the obvious with magnificent hyperbole in his "Book of the Machines": "Man's very soul is due to the machines: it is a machine-made thing: he thinks as he thinks, and feels as he feels, through the work that machines have wrought upon him, and their existence is quite as much a *sine qua non* for his, as for theirs" (*Erewhon*, 234). In the computer age, when hyperbole has become commonplace, man has no difficulty regarding even himself as an artifact, and with him nearly everything of interest in nature has been made over by technology.

This is the point of programs for artificial intelligence, and Herbert Simon has said as much in his book *The Sciences of the Artificial*. The evidence suggests, he writes soberly, "that there are only a few 'intrinsic' characteristics of the inner environment of thinking man that limit the adaptation of his thought to the shape of the problem environment. All else in his thinking and problem-solving behavior is artificial—is learned and is subject to improvement through the invention of improved designs" (*Sciences of the Artificial*, 26). Man is born an information processor with an empty memory store, and he programs himself to become an adult problem solver.

This new twist on Locke's old empiricism is the defining philosophy of the computer age. It certainly reflects the optimism of the Enlightenment. By ignoring the complex and hard-to-fathom aspects of human nature—what Simon calls in typical behaviorist fashion the "inner environment"—computer man can bend himself to any task, find a rational path to any goal. Surely human beings are not as flexible as that or as artificial. Recent hardwon knowledge in neurology and genetics shows us how little control we still have over those aspects of nature that most immediately affect our thought and action. It may be hundreds of years before man can really make himself over genetically, if ever. But Turing's man is so caught up in the computer metaphor that he refuses to wait for the genetic solution; he chooses to regard man as software more than hardware, as the program run by the computer more than the hard-wired machine itself. So he speaks of programming techniques (means-ends analysis, simulation, optimization, and so on) just as Enlightenment figures spoke of reason. The mind programs itself and through programming solves problems, achieves goals, molds itself to its environment. The making of man into an electronic artifact goes beyond the dreams of any nineteenth-century entrepreneur or materialist philosopher. For Turing's man, the ancient rift between the human and the natural is mended in a startlingly different way. It is not that man is a part of nature so much as the reverse—nature and man are both artifacts. For what is nature but a brilliantly designed "ecosystem," whose beauty and significance for mankind lies in its operational success?

In going to such extremes, Turing's man parts company with his predecessor. His concern with functions, paths, and goals overrides an interest in any deeper kind of understanding. In general, men of the computer age seem destined to lose the Faustian concern with depth. The rejection of depth for considerations of surface and form, long a feature of modern art, is now spreading throughout our intellectual life. The so-called sciences of human behavior make it their creed not to look at human experience below the surface. Indeed, they sometimes deny the existence of any experience that is not immediately and superficially demonstrable. Behaviorist psychology regards a man as a complex of sensing and responding elements that are wired together to produce human action; there is no question of deep, perhaps unfathomable motives and unconscious thoughts. Sociologists treat aggregates of human beings in the same operational terms, and economists treat them as unambivalent pleasure machines. Although computer technology did not single-handedly call forth this view, which has been developing at least since the turn of the century, it has nonetheless encouraged it, providing by far the most compelling metaphor in the social scientists' vocabulary.

Turing's men are by no means all strict behaviorists. Noam Chomsky is famous for his attack on the behaviorist view of language as a simple matter of stimulus and response. In fact, many programmers in artificial intelligence regard their work as "humanistic" psychology because their programs are meant to simulate the human mind as it manipulates symbols in sophisticated ways, not to mimic the simple reactions of chains of neurons. But programmers with their semantic networks and behaviorists with their Skinner boxes agree on this vital point: everything that happens in the mind or the brain is played out according to the rules of a formal system. These rules are finite, and they can someday be specified. Douglas Hofstadter, one of the most thoughtful of Turing's men, has made this a central theme of his book *Gödel, Escher, Bach*. For him the paradox is that the brain is a formal system of neurons underlying an apparently informal system, "which can, for instance, make puns, discover patterns, forget names, make awful blunders in chess, and so forth" (*Gödel, Es-*

cher, Bach, 559). Here is the classic philosophical problem of
mind and matter expressed in a way practically incomprehensible
to anyone before the twentieth century. To a Western European
metaphysician or a Platonist, it would have seemed an utterly superficial approach to the problem. How could a mere set of formal rules underlie the human intellect, with its access to universal notions of truth and divinity? Yet Marvin Minsky, the artificial
intelligence specialist, writes: "To me 'intelligence' seems to denote little more than the complex of performances which we happen to respect, but do not understand. So it is, usually, with the
question of 'depth' in mathematics. Once the proof of a theorem
is really understood its content seems to become trivial" ("Steps
toward Artificial Intelligence," 27).

The goal of artificial intelligence is to demonstrate that man is
all surface, that there is nothing dark or mysterious in the human
condition, nothing that cannot be lit by the even light of operational analysis. Like any program, an artificial intelligence program is a set of instructions to manipulate symbolic data: every
symbol and every instruction is as clearly defined and accessible
as the next. There are no shades or degrees, and nothing can remain undefined. A dislike of mystery is ingrained in every programmer by hard experience; for every one has spent untold
hours "debugging" his programs, tracking down subtle errors
that have crept into his commands as he wrote or copied them.
Unexplained or unknown lines of code do not add variety or give
his work a pleasantly unpredictable turn; they simply mean failure to perform. It is no surprise, then, that Minsky claims: "It
may be so with *man*, as with *machine*, that, when we understand
finally the structure and program, the feeling of mystery (and
self-approbation) will weaken" ("Steps toward Artificial Intelligence," 27). To put it another way, the symbolic logic by which
the machine functions demands total unidimensional understanding. The goal of logicians at least since Leibniz has been to shine
the light of mathematical reason upon the widest possible area of
human experience. Artificial intelligence programmers have pursued that end further than even Leibniz envisioned, for they even
devise algorithms to imitate human paranoia, thus reducing the
irrational to a set of machine instructions.

In his own way, computer man retains and even extends the
Faustian tendency to analyze. Yet the goal of Faustian analysis
was to understand, to "get to the bottom" of a problem; it divided
an issue painstakingly into parts in order to build a clear picture

of the interrelations. Turing's man analyzes not primarily to understand but to act. A computer program is not a static description but a series of instructions. This we have seen all along—a program is a logical theorem that proves itself by its execution. The computer gives mathematical and verbal symbols a life of their own, sets them dancing to a prearranged tune, and the programmer is never sure that the tune is correct until he can witness the dance. For Turing's man, knowledge is a process, a skill. A man or a computer knows something only if he or it can produce the right answer when asked the right question. The approach to any problem is still highly analytical but utterly superficial, for depth in the Faustian sense adds nothing to a program's operational success. Electronic man creates convenient hierarchies of action by dividing tasks into subtasks, routines into subroutines. The end is reached when the "subproblems" become trivial manipulations of data that are clear at a glance. In this way, all complexity is drained from a problem, and mystery and depth vanish, defined out of existence by the programmer's operational cast of thought.

Structural analysis is nothing new: think of Kant's elaborate hierarchy of human knowledge and experience in the *Critique of Pure Reason*. What is new is the enthusiasm for such structures, the conviction that everything we know (and perhaps all that exists) can be reduced to elements no more complicated than nodes in a tree diagram. That conviction surely lies behind the amazing behaviorist contention that men are mechanisms capable of only a few simple responses. Again Herbert Simon writes: "An ant, viewed as a behaving system, is quite simple. The apparent complexity of its behavior over time is largely a reflection of the complexity of the environment in which it finds itself. . . . I should like to explore this hypothesis, but with the word 'man' substituted for 'ant' " (*Sciences of the Artificial*, 24, 25). The choice of the ant is revealing, for an ant, unlike a dog or a cat, lives an existence that is thoroughly structured and entirely defined by trivial responses. Ants do not ruminate, cry, panic, or show fear or contentment; they act. Their actions are empty of content and take place within a community defined only by its geometric patterns of living production units.

The computer view likewise reduces much of human experience to geometric structures. The computer itself, so powerful at mathematics and logic, is a complexly linked matrix of simple electronic components. Is it not a model for the world in which it functions with such efficacy? And does it not permit computer man to mold even space and time into a structure to meet his needs?

Turing's man treats ideas in plastic terms: he shapes and re-shapes them very much as a child molds figures in a sandbox. To the previous age, most goals worthy of achieving and most important answers seemed difficult and remote. Goethe's character Faust was indeed an archetype of Western values in this respect, for the knowledge to which Faust aspired demanded years of study and a pact with the devil. Even then it could not be handed over. Faust could learn only through long and painful personal effort, through a spiritual and physical journey, a quest. The quest was the central theme of Western European literature. The medieval romance with its search for the Grail began a tradition that is not defunct even today. The alienated hero, the indefatigable star of movies and fiction since the First World War, is still on a quest for self-knowledge. Meanwhile, dynamic technology since the Middle Ages has been a quest for power, an obsessed effort from which there was no turning back. Western men and women tackled all problems, literary or technological, with the same determination to pursue the ever-receding limit of the ultimate. Even the businessman of the nineteenth and twentieth century, who has pursued wealth beyond all proportion, sometimes to the ruin of his health and family, is an expression of the same Faustian urge.

The computer man chooses another way. He remembers that computer programming is a game, the spinning out of solutions to well-defined problems according to strict rules. It is not a search for something remote, hidden, deep. A game is played with materials ready at hand; it may indeed be tricky or taxing but always within a familiar field of play. In most games, the elements are simple and are repeated from one playing to the next; what is novel is the way the elements fit together. Certainly this is true of electronic gaming, as well known to engineers who solve differential equations by computer as to amateurs who play Star Trek on their home machines.

The man on a quest had to go to the ends of the world to learn his difficult lesson; in general, he changed in the process. Not so for the computer programmer. He remains in the confined logical universe of his machine, rearranging the elements of that universe to suit the current problem. The programmer remains the same, and the world changes around him. Self-knowledge is not particularly his goal; self-improvement may be a goal, but this is understood in practical terms as increased efficacy. The programmer reworks his logical world to make it more efficacious or more comfortable, and he proceeds until he comes up against the ultimate electronic limitations of time, space, or logic. In the process, he learns nothing more than he put there himself, for he does not discover his world so much as invent it.

By its very definition, this game of invention is not as serious as a quest. The quester has more to lose, and perhaps more to gain, because the changes he undergoes are largely irrevocable. He succeeds or fails on a grand scale. Goethe's Faust put nothing less than his soul into the balance in his quest for knowledge and activity. It was characteristic of Western man never to be satisfied with anything less than the ultimate achievement and so always to remain unsatisfied. The explosive technology of the past six centuries has likewise been propelled by a perpetual dissatisfaction with the current ability to control nature, an endless search for a more powerful prime mover, a stronger metal alloy, a more precise mechanism.

Turing's man lacks the emotional intensity of his predecessor. He invests less of himself in his games precisely because the games he plays are not irrevocable. They are meant to be played to a conclusion and then reset and played again. The programmer indeed cares about the game's outcome, but he is saved from ultimate failure by its impermanence. A computer program that fails can usually be corrected and rerun. In another sense, this is a disadvantage, for a programmer can never forget that every solution in the computer world is temporary, makeshift, obsolescent. The electronic medium is in fact less permanent than even the sandbox to which I have compared it; the ease of transferring data and the huge amounts of data transferred mean that nothing in the computer world remains long in one form. Every programmer knows how easy it is inadvertently to destroy hours of work with the touch of the wrong button. But no program is intended to last long anyway. If it is useful and often used, then someone will soon modify it for his own plans, which differ slightly from the original purpose. Eventually it will be thrown out altogether, when new equipment or software makes it obsolete. A philosopher or poet of the nineteenth century might aspire to be read for hundreds of years to come. An inventor might hope to make a machine whose utility would extend for decades. A computer programmer or engineer measures his achievement in years, often months.

Lacking the intensity of the mechanical-dynamic age, the computer age may in fact not produce individuals capable of great good or great evil. Turing's man is not a possessed soul, as Faustian man so often was. He does not hold himself and his world in such deadly earnest; he does not speak of "destiny" but rather of "options." And if the computer age does not produce a Michelangelo and a Goethe, it is perhaps less likely to produce a Hitler or even a Napoleon. The totalitarian figures were men who could focus the Faustian commitment of will for their own ends. What if the will is lacking? The premise of Orwell's *1984* was the marriage of totalitarian purpose with modern technology. But the most modern technology, computer technology, may well be incompatible with the totalitarian monster, at least in its classic form. Nazi Germany was technologically deeply committed to the mechanical-dynamic model: vast sources of power were devoted to military problems, and everything from tank warfare to architecture was done on the largest possible scale. It is no accident that the autocratic regimes of Eastern Europe today are almost untouched by the computer age. Computers make hierarchical communication and control far easier, but they also work against the fundamental sense of purpose, the absolute dedication to the party line, which is the core of the autocratic state. The computer programmer is always aware of other options. If anything, the great political danger of the computer age is a new definition of anarchy.

That danger is perhaps all the greater because the computer man is less aware of history than his predecessor and is not likely to see the historical currents in which he is caught. He tends to project the present indefinitely in both directions. As stated before, one obvious reason is the rapid development of the technology with which he works. Turing's man is in one sense a great believer in progress, a belief that ought to give him an interest in history. His predecessors in the eighteenth century combined a belief in human progress with a strong historical sense. In fact, they initiated modern historical writing. The irony is that progress in the computers and in the rest of our current technology is so rapid that it tends to negate history. In the past, technological progress was a matter of years and decades; even the engineer had to take a somewhat longer view of his subject. Today engineers, like the Red Queen in *Through the Looking Glass*, must run as fast as they can to remain in one place. They can hardly afford to look very far back, for in high technology anything more than a few years old is obsolete. They generally look ahead only as far as the presently understood technology can lead them. And so they live on an island of time, whose boundaries are the few years on either side of the technological moment.

In any case, computer man is not likely to appreciate deep-seated motives that may propel individuals and whole societies along a particular path. The notion of the driven man or the society that develops organically according to some overriding principle was one that appealed to Faustian man. When he thinks of history at all, Turing's man projects his own technological frame of mind far into the past; history is for him the development of techniques for the manipulation of nature, an evolution that would have proceeded steadily if scientists and craftsmen of the past had not been opposed by the ambitions of rulers and the superstitions of their uninformed or greedy subjects. In fact, the attitude toward history fostered by electronic technology is like that of the ancient Greeks, who lived in the present both in their politics and in their literature. But if computer man, like ancient man, lacks the historical perspective that we prize, he has the advantage of being free to some degree of the weight of tradition, making it easier for him to operate within the intellectual world of his own making.

Turing's man does feel limits, not the confines of tradition but instead the ultimate confines of his materials. Perhaps the most revolutionary change of all is that the computer man thinks of his world, intellectual and physical, as finite. He has no sympathy for the endless striving in an endless universe that characterized his predecessor. He does not worship the infinite; he does not view it as heroic to attempt the impossible. If we count the ways that the Western love of infinity has shaped our culture, we begin to grasp the dimensions of the change that is in store. For example, the Western concept of God as an infinite being must surely fade. The ancients did not think of their gods as infinite, nor did they generally think of infinity as a worthy attribute. Western theologians, however, found it their main task to explain how an infinitely wise, good, and powerful deity could have anything to do with a finite creature such as man and with the imperfections of nature as man knew it. If religion survives in the computer age, it will do so only by accommodating itself to a new, drastically re-

duced scale for both God and nature. The arts have already made
such accommodations. The plastic arts and music (especially
electronic music) have been developing for decades in the direction of the immediate and the finite. The rejection of linear perspective in art, which occurred long before the development
of the digital computer, fits well into the electronic scheme of
things, and the emphasis in recent music on spinning out compositions from a finite set of ready-to-hand materials is also in the
spirit of the computer age. How literary forms will change remains unclear.

The case of social organization is clear. Mechanical-dynamic
technology led mankind to pursue the politics and economics of
infinity, a policy now being challenged by a new technological
outlook. The global turmoil we are now witnessing largely results from the inability of peoples and governments to break with
their old technological faith. In the old view, the entrepreneur
was a Faustian figure with an insatiable desire to control nature.
And if nature could never be completely dominated, completely
transformed into capital, that too was the glory of the entrepreneur—his work was never done. Mechanisms must always be
made more exact, metals converted into stronger alloys, new
sources of power exploited on a grander scale. The national or
world economy would grow forever as raw materials were
formed into ever more sophisticated products, and national popu-
lations would continue to expand to provide the needed labor.
This view, so successful in the last two hundred years, still pre-
vails in all modern societies, leading to the absurd excesses of
population and consumption that now threaten us.

The computer was born in the final and most spendthrift decades of the Western economic growth. Yet by its very nature it
encourages a finite world view, and this may well be the greatest
good fostered by the computer. The prime task of the programmer is to manage his scarce resources, to accomplish what he can
with near and ready materials rather than solve problems by ex-
pansion. Computer specialists are always looking forward to expanded systems—more memory and faster central processors—
but they know that no improvement will remove the finitude of
their machine. It is no accident, then, that computer control and
simulation figure so largely in attempts by government and industry to manage shrinking resources: the computer was made for
such work, just as the steam engine was perfectly suited to the
expanding national and imperial economies of the last century.

For those who find that the evils of this computerized world view outweigh the benefits, who think that the superficial attitudes toward human and social motives and the lack of historical sophistication are more important than the new economic attitudes, I offer the following consolation: these are the tendencies, the preferences of Turing's man; they need not overwhelm us. In fact, these very tendencies are likely to provoke a vigorous reaction in our cultural life. After all, the romantic and naturalistic movements of the eighteenth and nineteenth centuries developed as a response to the Industrial Revolution and the social changes it brought. The sensitivity to nature that many now regard as a human instinct was in fact the work of poets and philosophers—Rousseau, Blake, Goethe, Wordsworth, and others. The Romantics found industrialization and complex urban life appalling because of its misuse of nature, its attitude of exploitation, but it was that very attitude against which they reacted so creatively.

Might not our reaction to distasteful aspects of the computer age be equally creative? If Turing's man is unaware of history, it is the responsibility of writers and artists to educate him otherwise. If the computer programmer is more isolated from nature than even the industrial worker, humanists should define a new relationship to nature, drawing on the best qualities of electronic technology. (I come once more to the computer as a game: May we not find ourselves playing a game with nature rather than struggling against it?) Likewise, computer man's oblivion to deep human motives should provoke us to redefine and emphasize just this aspect of the human condition. Where the computer sees human nature, like every other field of information, as a rigid hierarchy of data and control, we should explore the unstructured and spontaneous elements of human action and intention.

The success of this reaction will depend not only upon programmers but also upon humanists, who will have to learn about computer technology in order to trim its excesses. Decades of novels, movies, and essays on the theme of "alienation in a technological society" show how barren the reaction may be if it is not informed by some sympathy for and knowledge of its proposed subject, technology. Humanists must realize too that they were nurtured on nineteenth- and early twentieth-century values, not all of which can be preserved for the very different future we now confront.

Cultures change in part because of a momentum they have acquired over decades and centuries; we can seldom nullify the momentum or reverse the direction of cultural change. We can perhaps bend the course to make it more to our liking. But we should not be sanguine about applying external controls to technological change; it is seldom effective to ban research or forbid uses of the computer of which we do not approve. Such controls usually fail. In England the struggle against the mechanization of textile production was violent and sometimes bloody, but in the end it hardly succeeded in postponing the inevitable. Clearly we must sometimes resort to political remedies; we must legislate against computerized spying and record keeping by business and government, for example. But we should view such legislation as a last resort, as an admission that our culture has failed once again to make wise use of a technological advance and so must act to prevent its misuse. By the same token, refusing to work on some electronics project may be the best ethical course for an individual engineer, but it does little for our society as a whole. An engineer who refuses to build microprocessors for cruise missiles may earn our admiration, yet his refusal will not affect the arms race. Others will be ready to take his place. A far better strategy is to work at reforming the age of computers from within by finding compelling uses for computers that respect the differences between men and machines. In this way, we can build into the machine itself a bias for the humane treatment of human beings. And this task calls in particular for those who feel most isolated in the electronic age: the artist, the poet, the historian, the humanist.

Let me be specific. Turing's man is insensitive to the historical and intellectual context for his work. He tends to see the past as an indefinite extension of the technological present. We must find ways to bring history into the computer world. At the very least, we must make sure that the record of history and culture is preserved in the coming age. It is a truism but nonetheless true that we are about to witness a great transfer of knowledge from one medium to another. For hundreds of years, books were the principal means of preserving and spreading our collective wisdom. Now books are being challenged for that honor by electronic storage devices; data bases are challenging libraries. There are at least two historical parallels: the change from manuscript to print in the fifteenth century and the change from uncial (capital) to minuscule script in ninth-century Byzantium. In both cases, the new medium destroyed the old. Indeed, in their eagerness to get

out a printed edition, editors would buy a manuscript and tear it
to pieces as they set their type. Some manuscripts valuable to pa-
leographers were lost in this way. In Byzantium ancient texts that
were not copied into the new script did not survive, and many
ancient Greek authors were lost forever.

I do not care to argue that the computer will likewise make the
book obsolete; printed books are portable, convenient, and dura-
ble. However, it is terribly important that the knowledge we now
keep in books be transferred to the electronic medium, where engineers, scientists, and public figures will be doing more and
more of their work. Until now, data bases have been ruthlessly
practical. Most of them store business information; some preserve the swelling bibliographies of science and medicine. Probably the most wide-ranging ones at present are the *New York Times*
News Service (containing current articles) and the legal data base
of Supreme Court reports. Admittedly, the first data bases had to
have some economic reason for existence. But now that the theory has been developed and storage devices are becoming more
capacious and less expensive, it is possible to satisfy other needs
than the purely practical.

In the eighteenth century, Diderot's great *Encyclopédie* was an
expression of the whole age. It was eminently technological, including eleven volumes of illustrations of the crafts of the day,
yet it also gave voice to the humane and highly political sen-
timents of the leading thinkers. The time has come for an electronic encyclopedia that would be a similar marriage of technology and humanism. Students of various languages and literatures,
historians, critics of art and music, philosophers, sociologists,
and anthropologists could all take part. Eventually, vast amounts
of literature and history could be made available to the user—to
study one area intensely, to search many authors, genres, or periods for passages appropriate to his subject, or simply to browse.
The browsing would be motivated by the same disinterested curi-
osity that is now possible in a library, but the electronic medium
would open up new ways to look, suggest new questions, and
make at least some new answers possible. For the technique by
which the computer organizes knowledge—nuggets of information linked together into a complex and flexible structure—would
make a data base of humanistic literature very different from a
library. It would allow the reader to treat written knowledge as a
vast workspace in which to build his own interpretive structures.

Something would indeed be lost by this transfer to a new medium: the impressive sense of permanence that printed books and massive library buildings convey. The fixity of the text and the glory of lasting literary monuments would be diminished. In their place would be the qualities of the computer age, namely, the sense that recorded knowledge can be collected, transmitted, reorganized, and molded to one's immediate needs and the feeling of being immediately "in touch with" our culture's past. The technologist might find the past more accessible in this form, and the humanist might find new uses and new interpretations for familiar texts and historical events.

Invention and Discovery

The computer is often blamed for destroying individual freedom, the right to be different. It is true that the computer does not lend support to the eighteenth-century notion of a free individual who answers only to his own reason in deciding his destiny. But the real culprit here is certainly social forces outside of the electronic pale. Chief among them is the overpopulation that leads to powerful and distant government and in general diminishes the value of any one person. On the other hand, the computer does suggest a new way of looking at individual freedom and creativity.

In the previous age, the individual was regarded as a self-sufficient ego, an ego moving out and acting in the infinite world of society and nature. The individual might well confront and conflict with other egos in his journey. But in an unlimited environment, it was always possible to rebound in another direction. The purpose of this ceaseless activity was discovery, for there was a vast universe to be explored. In the computer age, the individual does not move out indefinitely from his starting point; he knows the limits of his working environment from the outset. The computer is generally a crowded place, and around each user on all sides are others with whom he must share the resources of the machine. In any case, the programmer's task is not so much to discover but rather to invent. His workspace gives him limited resources, but resources nonetheless, and indeed even the limitations become a subject of his inventiveness. Working with a computer, the individual asserts himself not in the spirit of domination of ever-expanding realms but in a more measured way. This is a paradigm for individual expression in a crowded world.

The computer shows that even teamwork need not thoroughly

subsume and homogenize the special contribution of each member. The best organization for many computer projects is modular: each member of the group is given a separate part of the larger program or machine design. This is not the stultifying specialization of the assembly line, where one worker performs one operation repeatedly for hours. Instead, each module may be a self-contained program or portion of hardware, with challenges and difficulties all its own. The relationships among the various modules is specified by the team: what the module will receive as input and what it must produce as output. The programmer is left free to meet these specifications as he sees fit; the module belongs to him and is an expression of his prerogatives, even of his personality. He is no longer the thoroughly unfettered inventor of the nineteenth century, who could work alone in his shop, drop one project and start another, or change his mind completely about what he was trying to invent. But he is no mere automaton; he can work with some spirit and self-esteem. And he is likely to appreciate the necessity of working with others.

I am probably putting the case too mildly, for computer specialists do not seem to feel hemmed in by teamwork and limited resources. They are often exhilarated by the work, as is indicated by Tracy Kidder's book on the design of a new minicomputer, *The Soul of a New Machine*. These designers worked passionately for months, sometimes seven days a week, in their effort to tinker together a new computer. The joy of the work and friendship and rivalry within the team drove them on; money was clearly of secondary importance. Most of them seemed to value the fact that some particular system or aspect of the computer was their own doing. Kidder also makes clear that this interplay of ambitions and talents could only last so long under the pressure of "getting the machine out the door." Teamwork and modules are not the ultimate answer to the conflict of individual will and the collective good, but they suggest a working compromise for the coming era.

The Computer as a Tool

The computer is in some ways a grand machine in the Western mechanical-dynamic tradition and in other ways a tool-in-hand from the ancient craft tradition. The best way to encourage the

humane use of computers is to emphasize, where possible, the
second heritage over the first, the tool over the machine.

A machine is characterized by sustained, autonomous action. It is set up by human hands and then is more or less set loose from human control. It is designed to come between man and nature, to affect the natural world without requiring or indeed allowing humans to come into contact with it. Such is the clock, which abstracts the measurement of time from the sun and the stars; such is the steam engine, which turns coal into power to move ships or pump water without the intervention of human muscles. A tool, unlike a machine, is not self-sufficient or autonomous in action. It requires the skill of a craftsman and, when handled with skill, permits him to reshape the world in his way. For the Greek craftsman, the tool was always either an extension of the hand (a last or a hammer or a brush) or a more complicated arrangement that presented materials to his hand for shaping (a loom or a potter's wheel).

However, the computer is not really a tool-in-hand; it is designed to extend the human brain rather than the hand, to allow the manipulation of mathematical and logical symbols at high speeds. Yet it can be used with a kind of mental dexterity and reminds us of the craftsman's hand. it is precisely this analogy that is expressed visually in an early scene of Kubrick's film *2001* —when the hominid in triumph whirls the first grasped tool, a bone, into the air, and it instantly becomes a rotating space station. By leaping over a million years in between, the camera equates the first tool with the last. And we soon realize that this last tool of interest is not the station itself but the electronic logic machines that control space vehicles in the year 2001.

To use the computer as a machine is to emphasize its admittedly great autonomy from human control, to explore ways in which it can remove humans from contact with the world around them. To use the computer as a tool is to give the man with a "computer-in-hand" a more effective grasp of his physical and intellectual milieu. When I claim that the artificial intelligence movement illustrates both the danger and potential of the computer, I have this dichotomy in mind. No one has pursued the computer as machine more doggedly than the artificial intelligence specialist who has tried to replace human thinking with a computer algorithm, construct a problem solver that will supersede human powers of reason, or design a communications program to "process" natural language without human intervention. Such projects do have demonstrative value, for they impress us with the new qualities of the computer as a symbol manipulator and a logical agent. But they do seem to violate an obvious division of labor: to let the computer do what it does well and to allow humans to intervene where their talents are called for. This principle, of course, is not new; many computer specialists have appealed to it before. Walter Rosenblith, who has studied both medicine and the computer, remarked two decades ago: "I am . . . less tempted to stress what computer can do better than men than to envisage the benefits that we might derive from an intelligent division of labor between man and computer." He then continued: "The combination of man *and* computer is capable of accomplishing things that neither of them can do alone" (cited in *Computers and the World of the Future*, ed. Martin Greenberger, 312, 313). And John Kemeny, the educator and computer specialist, has commented: "One can't help reaching the conclusion that it is more efficient to use a human being as the computer's partner than to spend many years trying to teach a computer a talent for which it is not well suited" (*Man and Computer*, 17). It is not merely a question of efficiency, of course. The demeaning and dangerous aspects of the computer age come precisely from this failure to respect human talents. Fortunately, efficiency and humane good sense are allies in this instance.

When the mathematicians Appel and Haken proved the four-color-map theorem, they used the computer as a logical tool. They did the analytic work of reducing the scope of the problem to an electronically manageable size and then called on the computer to check through hundreds of maps to complete the proof; mathematicians had been working for a century on the problem without being able to see it through to the end. This illustrates the use of the computer as an extension of the human ability to reason logically. Used as a wholly autonomous logic machine, the computer is much less effective. With great programming skill, it has been coaxed into proving some trivial theorems in Euclidean geometry and simple results in symbolic logic.

To take another example, why seek to eliminate the human element from the experience of language? In the fifties and sixties, at great expense to the government, programmers worked to replace the human translator with a machine. Their projects betrayed Americans' distaste for learning foreign languages; the hope was that the computer would learn Russian so that American scientists would not need to. They failed because the machine could not deal with the context of a word or sentence or with unusual grammatical constructions. Since then, artificial intelligence programmers have restricted themselves to tiny subsets of the English language in their effort to give the computer just this contextual understanding. The sample dialogue from SHRDLU in the previous chapter indicates how far they have gone, but the complexity of their programs makes it impossible to work the same magic over the whole of one language or over two languages for translation.

In the capacity of translator, the computer is a language machine. But suppose we treat it instead as a language tool. The computer is an ideal dictionary and reference grammar, for it can provide definitions and rules in an instant. We can allow our Russian program to look up words and translate phrases as best it can. When it gets stuck or clearly goes wrong, we let the human reader intervene. Such a program requires that the reader know a little Russian to begin with and expects that he will improve his knowledge as he goes. Gradually he will wean himself from the computer altogether, but in the meantime he will read far more enjoyably and quickly than he could without his helper. The principle here applies to many useful projects: build into the program a place for the human operator to intervene with his special insights into language, mathematics, medicine, and so on. The combination of man and computer will then have a greater range than either alone, and, just as important, such a combination will be more humane.

Let me return to the analogy between a hominid grasping his bone weapon and a homo sapiens seated at a computer terminal. The computer too is a prehensile tool, and it encourages an almost tactile approach to solving problems. Through its circuitry, the human operator "manipulates" his information, probes his files to examine their contents, builds data structures and tears them down, works and reworks input to generate output. This activity of give and take, especially on a modern interactive computing system, has much of the intimacy of the potter shaping his clay—all the more remarkable when we remember how abstract this electronic clay really is. Even time and space can now be reworked and reshaped by the programmer. It is nothing new to use metaphors of touch to describe mental effort—to grasp an issue, comprehend a subtlety—but such metaphors were never more appropriate than they are for the computer age.

Since the invention of printing, European culture has valued the sense of sight over that of touch. After all, seeing is believing. Seeing encourages analysis, theory, and metaphysics, all of which have flourished for the last several centuries. But the computer, the product of solid-state physics and high technology, is nonetheless an eminently practical machine, and it encourages Turing's man to take a pragmatic approach to the problems he encounters. The notion of grasping a problem, of trying out solutions by manipulating a "solution space," challenges the primacy of theoretical, visual analysis. Computer simulation offers instead the pragmatic method of plugging in sample values and letting the program run. This method of trial and error resembles nothing so much as the craftsman testing his work by hand, feeling for the desired shape.

In fact, the programmer at his terminal is no less removed from the world than the nineteenth-century mathematician in his study, his program no less an abstraction than the mathematician's equations on paper. Yet the feeling of the work is different. It is no longer deep analysis but the practical solution that counts. Unquestionably the old attitude toward intellect had great beauty, but the tactile approach of computer simulation is far more accessible. If the programmer does not need fine mathematical skills or powers of deduction, he nonetheless finds satisfaction in the extension of mental grasp that the computer allows. Here is the computer's great potential: to provide prearranged elements that the programmer can assemble in imaginative ways. The machine's automatism becomes the foundation for the greater imaginative freedom of its operators.

By making complex techniques available in a simplified form, the computer encourages a new kind of amateurism. As we have seen, the notion of hierarchy is fundamental to a computer system. The user can work at one level without understanding the details of operation at the levels below him. He does not need to understand solid-state physics or assembly language in order to write programs in a high-level language such as PASCAL. Hierarchical construction means that any one programmer can rely upon the collective effort of dozens or hundreds of others who have worried over details that no longer concern him. He can thus invoke mathematical programs simply by knowing how to express the input and how to interpret the output. In this way, the results of sophisticated mathematics, though not the deep analysis, become available to millions of amateurs.

The benefits of hierarchy are not limited to mathematics, however. In all problems that require the manipulation of symbols, amateurs can operate practically, though not theoretically, at the level of experts. And in an age of specialization, this is a healthy sign. Might it not lead to a cultural preference for experimenting, playing, and perhaps even browsing in areas where one has not got some officially certified expertise? The age of the expert is, of course, quite recent. Even as late as the eighteenth century, no stigma was attached to being an amateur who was motivated by love of the subject rather than professional obligation. The computer will in no way eliminate the need for thorough knowledge and fields of specialization, but it might mean that the specialist need not be banished from a lively interest in both science and culture at large.

It may even help to return creativity to the sphere of the amateur. Today the arts are marked by the same specialization as the sciences. The music of the nineteenth century demands virtuoso skill; the music of the twentieth demands the same skill as well as a finely schooled sensibility. And the best contemporary music is inaccessible except to a few who have taken years to become acquainted with its aesthetics. Much the same is true of the fine arts and to a lesser extent of literature. The computer offers at least the possibility that the creative process may be made hierarchical as well.

A composer-programmer may create a system for generating electronic music and express it in a simplified programming language. An amateur who himself does not know the details of sine waves and sound envelopes may then use the program to generate his own compositions. One can imagine every degree of interplay between the composer and the amateur: the composer may exercise great control over the sound production and leave the amateur to manipulate relatively few parameters, or the amateur may have great scope for making his own sounds. But in any case, the amateur would participate much more actively in the making of music than he does today as a passive listener at a concert. The creative act would then be an interplay between the composer, who sets up the framework and defines the limits of the composition, and the amateur, who realizes one of the compositions that the composer has made possible. (This is not far from the practice of many recent composers working with traditional and sometimes electronic instruments.) Graphic display would allow artist and amateur to play a similar game in the visual realm.

Even literature might yield to this hierarchical approach. The key is to remember to regard the computer as a medium for expression, not as the creator itself. By storing, organizing, and realizing the work of the programmer, it becomes a creative tool, not a creation machine.

Synthetic Intelligence

In promoting the computer as a tool, I am arguing for a synthesis of man and computer, rather than the replacement of man by machine. It seems to me that "synthetic intelligence" would be a happier name and a better goal than "artificial intelligence." Nor are the programmers of the artificial intelligence movement always far from the mark. They do indeed express the promise as well as the dangers of the computer age, and some of their best work is directed toward making tools for the creative interaction of humans and computers.

I think in particular of an educational programming language called LOGO, which was designed by Seymour Papert and described in his book *Mindstorms*. His idea was to allow even young children to learn geometry and mathematics actively by programming a computer. The children specify geometric designs in simple commands, and the computer presents them with a visual realization of their efforts. The designs are in fact drawn by a "Turtle," which may appear on a television screen or may be a mechanical robot that makes lines on a piece of paper on the floor. In either case, the children work in active partnership with the computer. In ordinary computer-aided instruction, says Papert: "The *computer is being used to program* the child. In my vision, *the child programs the computer* and, in doing so, both acquires a sense of mastery over a piece of the most modern and powerful technology and establishes an intimate contact with some of the deepest ideas from science, from mathematics, and from the art of intellectual model building" (*Mindstorms*, 5). The choice of words here ("intimate contact with . . . ideas" and "intellectual model building") show that Papert is not trying to eliminate the computer's suggestive influence upon the child's thinking. Indeed, he calls his Turtle an "object-to-think-with."

What sort of thinking will this object encourage? The older, Western style of analysis will be translated into simulation, the new electronic style of problem solving. If the child wants to draw a flower, he writes a program to direct the Turtle. If the petals come out looking like fishes, he refines his program. Gradually, after many attempts, the flower takes shape, and soon it grows into a garden. Children playing with the Turtle are learning to think in a way suitable for the computer age: by trial and error, step by step, with a feeling for the spatial limitations of their diagrams. Their programs are acts of synthetic intelligence; they combine electronic graphics and logic with the children's own motives and insights. LOGO will not encourage children to become mathematical analysts, philosophers, or authors in the nineteenth-century mold. Yet they may become equally creative in a new style of thought—computer game players at their best.

Glossary of Computer Terms

What follows is a devil's dictionary of the computer, that is, I have incorporated my own arguments in many of the definitions. The result is a list that should be used in connection with this book and not as a general reference. A good reference work for laymen is the *Encyclopedia of Computer Science* by Anthony Ralston and C. L. Meek (New York: Petrocelli/Charter, 1976).

Address. The number assigned to each storage location in the computer's *memory*. The address allows a program immediate access to any location and makes possible the mathematization of memory. Addresses serve as pointers or links among storage locations, and these links can be used to build *trees*, arrays, and other multidimensional structures.

Address space. The numbered cells of computer *memory* that are available to a program for fetching and storing data. The numbers or addresses allow the program to ignore the physical location of the data and build abstract structures in memory. Addresses are links between the *physical space* of the machine and the *logical space* created by a programmer to solve a specific problem.

Algorithm. A strategy for solving a problem that can be expressed in a *program* and so executed by a digital computer. Turing's men, particularly specialists in artificial intelligence, believe that human reason is essentially algorithmic.

ALU. An abbreviation for *arithmetic and logical unit.*

Arithmetic and logical unit. The portion of the central processor that operates upon small packets of data. (The rest of the *CPU* controls the movement of data from place to place.) The operations take the form of arithmetic or logical calculus, hence the name. The arithmetic and logical unit contains within it all the manipulative or data-processing power of the machine. The surprising fact is that a small repertoire of simple operations can apply to such a wide variety of programming problems. Abbreviated ALU.

Artificial intelligence movement. A group of computer specialists seeking to make the digital computer reflect the human capacity to reason. Its goal is to demonstrate that man should now be viewed as an "information processor."

Assembler. A computer program that translates from *assembly language* into *machine language.*

Assembly language. A code for programming the computer at one remove from the *machine language.* Assembly language is easier to use than machine language because it allows the programmer to substitute meaningful names for the arbitrary strings of 0s and 1s that drive the machine. Each computer or family of computers may have its own assembly language. A typical instruction in assembly language might be: "ADD SUM,INC"—add the value of the variable INC to the current SUM.

Bit. A binary unit of information, stored as a logical 1 or 0. Bits are represented in the processor generally as low or high voltages. The all-or-nothing quality of the bit affects every aspect of the machine's operation and therefore of Turing's man: all information processed in the machine must be represented as strings of discrete binary units.

Byte. A common measure of electronic information: 1 byte = 8 *bits*. A byte carries enough information to represent each of the letters, numerals, and control codes used by standard computers.

Central processing unit. The portion of the computer hardware that fetches and executes instructions and so processes data. It contains the *arithmetic and logical unit* as well as circuitry to decode instructions and move the data stored in *memory*. Abbreviated CPU.

Compiler. A program that translates from a *high-level language* (such as FORTRAN or PASCAL) into a lower level (such as an *assembly language* or *machine language*). A compiler makes it possible to write programs in a form more accessible to the human programmer. It mediates between the programmer's mathematical or logical symbolism and the 1s and 0s that the machine understands.

CPU. An abbreviation for *central processing unit*.

Data base. A complex arrangement of programs and data for storing and retrieving large amounts of information (generally today for business or library work). It occupies the highest level in the hierarchy of computer memory systems.

Decoder. The part of the *CPU* that converts machine instructions (written in binary code) into a sequence of electrical signals that control logical operations and the movement of data.

Formal system. A structure of symbols and logical rules. The rules are applied to the symbols in discrete steps to produce valid results—theorems of the system. A Turing machine is a formal system, and every computer is its embodiment. The question is whether human thought itself is the product of a formal system. Odd as it sounds, this may prove to be the great philosophical issue of the next few decades.

Hardware. The physical components of the computer, as opposed to the *software* or programs.

High-level language. A programming language like FORTRAN or PASCAL that allows the programmer to work with familiar mathematical or logical symbols. A *compiler* is needed to translate this language into *machine language* that the computer can execute. A typical instruction might be: "SUM = SUM + INC"—add the value of the variable INC to the current SUM.

Infinite loop. The bane of programming. In an infinite loop, the computer executes the same group of instructions repeatedly without accomplishing its task and terminating. The computer is helpless before anything infinite.

Leaf. A terminal *node* of a data structure called a *tree*: a node from which no further links arise.

Linear access. A method of getting at data on a storage device, such as a tape drive. With linear access, one must start at the beginning and read through to the desired portion of data. This is slower than *random access*, but it is characteristic of some devices, new and old. A book without a table of contents or an index would require linear access, that is, one would have to start at the beginning and leaf through the pages in order to find a particular passage. (For computer specialists, the preservation of knowledge in any form is a technology with such operational features as: speed of retrieval time, expense of storage per byte, method of access, and so on.)

Logical space. The computer's storage of data as seen by the programmer. Logical space is an abstract, geometrical, and mathematical field in which the programmer can build his data structures.

Loop. A group of programming instructions meant to be repeated many times on similar elements of data. The program loops through the data, putting out units of information, until it encounters some terminating condition and halts. If it never achieves the terminating condition, it is in an *infinite loop*.

Machine language. The binary code of instructions that the *CPU* can decode and execute. Other programming codes (such as *assembly language* and FORTRAN) must be translated into machine language for execution by the computer.

Man-month. Programming projects are often measured in man-months, the product of the number of programmers working by the number of months worked. The term illustrates the temptation on the part of Turing's man to mathematize human time. Human productivity and creativity cannot be rigidly measured by the simple scale of time logged on the machine.

Memory. In general, all the hardware and software devoted to the storage of data. Internal memory is that portion of the machine (now usually made of transistors) that preserves programs and data awaiting execution in the *CPU*. External memory is any storage device, such as a tape drive, that allows longer term preservation of information for possible use in the machine.

Microcomputer. A "desk-top" computer, generally built around one or a few microelectronic chips.

Minicomputer. Larger than a *microcomputer*, but smaller than the expensive "main-frame" computers. A minicomputer is used in business or scientific applications where great computing power is not needed.

Multiprogramming. The technique of allowing several programs to reside in a computer's memory at one time. The *CPU* devotes a fraction of its time, on a rotating basis, to each program. This kind of rationing of time and computing power is characteristic of the electronic age. It is the prototype of a new attitude toward the sharing of technological resources.

Natural language. Human language—English, Greek, Pali—in contrast to the codes used by programmers, which are called "artificial languages." The question is how natural language differs from the artificial codes; Turing's man is inclined to analyze natural language as if it were a code for storing and retrieving information.

Node. An elementary unit in a spatial data structure, such as a *tree*. The computer encourages us to represent knowledge and experience in terms of nodes and links—elementary data and their structural relationships.

Number-cruncher. A large, fast computer that is generally used for scientific purposes. The name suggests a mathematician's notion of power—the ability to work voraciously with numbers.

Numerical analysis. Computer mathematics involves the use of numerical techniques (complicated and subtle arrangements of arithmetic operations) to compute answers to problems of higher mathematics (integration, differential equations, and so on). The computer provides a remarkably successful but concrete and finite approach to mathematics.

Opcode. Part of the *machine language* of the computer: the system of 1s and 0s that tell the machine what operation (addition, subtraction, etc.) to perform.

Physical space. The numbered storage locations available in a computer's memory. Physical space is the stuff from which the programmer's *logical space* is in turn created.

Program. A sequence of instructions that can be executed by a computer, often synonymous with *algorithm*. More precisely, an algorithm is a statement of strategy, whereas a program is the expression of that strategy in a computer language such as PASCAL.

Random access. The fastest way of reading or writing data in computer storage. Random access allows the system to reach a particular portion of data directly, without considering any other data that may be stored in the same storage device. (A disk drive is a random-access storage device.) This method contrasts with *linear access*.

Real time. One of the more ironic terms in the vocabulary of computer specialists. Real time is time as the human programmer experiences it or as the outside world measures it. A computer system that works in real time must respond quickly enough to serve a human sitting at a console waiting for an answer or an industrial process waiting for control instructions. Putting a computer in charge of a power utility or an assembly line is an example of a real-time application. The system cannot go off, ruminate, and deliver a response the next day; the problem is in some sense too urgent. "Real time" is therefore the programmer's expression for an immediate interaction between the computer's microseconds and the needs of the outside world.

Software. Programs that the computer may execute, as opposed to the hardware or physical components of the machine.

Supercomputer. More or less the same as a *number-cruncher*; a large, fast computer used for scientific purposes.

Throughput. A measure of the speed and capacity of a computer—how many programs it can complete in a given period of time. An example of the mathematization of time encouraged by the computer.

Time-sharing. A method of dividing the attention of the *CPU* among several programmers; the CPU works for a moment on each program and then shifts to the next. The processor operates in its own temporal world, and often it is so fast that the user does not even notice that it is devoting only a small fraction of its attention to him.

Tree. A data structure consisting of links and *nodes*. The tree begins with a single parent node and grows downward; each node may in turn become a parent of several offspring nodes. Data are stored at the nodes of the tree, and the links indicate the relationships among the data. A tree is a common way of organizing knowledge in the computer world and shows how the programmer tends to express his problem in structural, spatial, and indeed geometrical terms.

Truth table. Used in logic to express all the possible "truth values" of an expression, based upon the truth or falsity of its constituent parts.

Turing machine. A rigorous logical definition of the operation of a computer. See chapter 3 for an explanation.

von Neumann computer. A design for a fully electronic computer drawn up by von Neumann and his colleagues after the Second World War. The design is still commonly followed today. See chapter 3 for details.

Wait state. The condition of the central processor when it is waiting for a signal to act upon a program. The computer does nothing profitable in the wait state; it simply loops until some user calls upon it.

Word. A unit of information larger than the *byte*. The length of a word varies from one computer to the next; it is commonly two or four bytes. Notice that *words* in the computer are not the same as words in human language: they are not nuggets of meaning but rather groups of binary digits that belong together logically and therefore travel together in the *CPU* and to and from *memory*.

Workspace. The particular area of storage given to a programmer during his active use of the computer. In his workspace, the programmer is free to build data structures and compose programs. He is not allowed to intrude into the workspace of others.

Annotated Bibliography

The following books and articles have been of particular value, either as sources for quotations or as background for the many subjects I have had to treat superficially in the course of my argument. Parenthetical references to the following works appear in the text.

Appel, Kenneth, and Haken, Wolfgang. "The Solution of the Four-Color-Map Problem." *Scientific American* (October 1977): 108–21.

Arbib, Michael A. *The Metaphorical Brain: An Introduction to Cybernetics as Artificial Intelligence and Brain Theory*. New York: John Wiley and Sons, 1972. Attempts, as the title suggests, to relate psychology and neurology by an explicit reference to the metaphor of the electronic brain.

Automatic Language Processing Advisory Committee, National Academy of Sciences, National Research Council. *Languages and Machines: Computers in Translation and Linguistics*. Washington, D.C.: National Research Council, 1966. The report that put an end to generous government support of programs for machine translation of human languages.

Ayer, A. J., ed. *Logical Positivism*. New York: Free Press, 1959. A convenient selection of positivist thinkers.

Boden, Margaret A. *Artificial Intelligence and Natural Man*. New York: Basic Books, 1977. A detailed account of numerous artificial intelligence programs, with an explanation of their significance. The author is clearly a partisan of the movement.

Boyer, Carl B. *A History of Mathematics*. New York: John Wiley and Sons, 1968.

——————. *The History of Calculus and Its Conceptual Development*. New York: Dover Publications, 1959.

Brooks, Frederick P. *The Mythical Man-Month*. Reading, Mass.: Addison-Wesley, 1975. An excellent account of the problems of programming teams. This book broaches the idea that the programmer creates from "pure thought-stuff," an idea I have found nowhere else.

Brooks, Frederick P., and Iverson, K. *Automatic Data Processing*. New York: John Wiley and Sons, 1963.

Bury, J. B. *The Idea of Progress in History*. New York: Macmillan Co., 1932. A survey that concentrates on the eighteenth century but has some interesting remarks on the ancient world in the introduction.

Butler, Samuel. *Erewhon and Erewhon Revisited*. New York: Modern
Library, 1927. A classic anti-utopian novel, yet Butler's views on
technology are much more complicated than is often supposed.

Cardwell, D. S. L. *Turning Points in Western Technology*. New York:
Neale Watson Academic Publications, 1972. A good, readable account of the clock and various heat engines of the eighteenth and
nineteenth centuries, as well as other technological breakthroughs.

Chapuis, Alfred, and Droz, Edmond. *Automata*. Translated by Alec
Reid. Neuchâtel: Editions du Griffon, 1958. Examples of mechanistic imitations of man, including the hydraulic and clockwork masterpieces of the sixteenth through eighteenth centuries.

Chomsky, Noam. *Aspects of the Theory of Syntax*. Cambridge, Mass.:
MIT Press, 1965.

——————. *Reflections on Language*. New York: Random House,
1975. Explores the broader implications of the new methods in
linguistics.

——————. *Syntactic Structures*. The Hague: Mouton and Co.,
1964. First published in 1957. By general agreement, a watershed in
modern linguistics.

Cipolla, Carlo M. *Clocks and Culture, 1300–1700*. New York: Walker
and Company, 1967. A good popular account.

Cohen, John. *Human Robots in Myth and Science*. London: George
Allen and Unwin, 1966. A fascinating story of mechanical and al-
chemical automata. It is quite useful for comparisons with the artificial intelligence project.

Cohen, Jonathan. "On the Project of a Universal Character." *Mind* 63
(1965): 49–63. Details this extraordinary caprice, which was popular in the seventeenth century. Leibniz was the most famous sup-
porter because of his proposal for a logical calculus of thought.

Collingwood, R. G. *The Idea of Nature*. Oxford: Clarendon Press,
1957.

Cornford, F. M. *Plato's Cosmology*. London: Harcourt Brace, 1937.
An excellent commentary on Plato's cosmological dialogue, the
Timaeus. Cornford recognized Plato's analogy between Greek
craftsmanship and the making of the cosmos.

Dales, R. C. "The De-animation of the Heavens in the Middle Ages."
Journal of the History of Ideas 41, no. 4 (October-December 1980):
531–50. Shows how medieval Europeans broke free of the ancient
concept that the stars are alive, with the help of comparisons between the heavens and such inanimate mechanisms as the mill and
the clock.

Descartes, René. *The Philosophical Works of Descartes*. 2 vols.
Translated by E. S. Haldane and G. R. T. Ross. Cambridge: Cambridge University Press, 1973–76.

——————. *Principia Philosophiae*. Vol. 8–1 of *Oeuvres de Des-*

cartes. Edited by Charles Adam and Paul Tannery. Paris: Librairie **249**
Philosophique J. Vrin, 1964.
Dijksterhuis, E. J. *The Mechanization of the World Picture*. Translated
by C. Dikshoorn. London: Oxford University Press, 1969. First
published in Dutch in 1950. A scholarly work which follows the
mechanical philosophy of science to its culmination with Huygens
and Leibniz.
Dodds, E. R. *The Ancient Concept of Progress, and Other Essays on
Greek Literature and Belief*. Oxford: Clarendon Press, 1973.
Dreyfus, Hubert L. *What Computers Can't Do: A Critique of Artificial
Reason*. New York: Harper and Row, 1972. This attack on the idea
of artificial intelligence takes a philosophical (phenomenological)
rather than broadly cultural or historical line.
Eames, Charles. *A Computer Perspective*. Cambridge, Mass.: Harvard
University Press, 1973. An illustrated essay based on an exhibit
by IBM.
Ebbinghaus, Hermann. *Memory*. Translated by H. A. Rogers and
Clara E. Bussenius. New York: Dover Publications, 1964. This
translation of the treatise *Über das Gedächtnis* (1885) has been
called the beginning of the modern psychological study of memory.
Edge, D. O. "Technological Metaphor." In *Meaning and Control: Es-
says in Social Aspects of Science and Technology*, edited by D. O.
Edge and J. N. Wolfe, pp. 31–59. London: Tavistock Publications,
1973.
Eisenstein, Elizabeth. *The Printing Press as an Agent of Change:
Communications and Cultural Transformations in Early-modern
Europe*. 2 vols. Cambridge: Cambridge University Press, 1979. A
thorough and lively account of this important subject.
Feigenbaum, E. A., and Feldman, Julian. *Computers and Thought*.
New York: McGraw-Hill, 1963. An early collection by the artificial
intelligence movement; contains Turing's paper on "Computing Ma-
chinery and Intelligence."
Finley, M. I. *The Ancient Economy*. London: Chatto and Windus,
1973.
Fishman, Katharine D. *The Computer Establishment*. New York:
Harper and Row, 1981. About the companies that make computers
and their own, occasionally appalling, world view.
Fodor, Jerry A. *The Language of Thought*. New York: Crowell, 1975.
A forthright statement of the electronic metaphor; for Fodor the human mind is without doubt an information processor.
Forbes, R. J. *Studies in Ancient Technology*. 9 vols. Leiden: E. J.
Brill, 1964– . The definitive work in English on the subject. Water
and wind power are discussed in volume 2, textile manufacture in
volume 4.

Fowler, Roger. *Understanding Language: An Introduction to Linguistics*. London: Routledge & Kegan Paul, 1974.

Giedion, Siegfried. *Mechanization Takes Command*. New York: W. W. Norton, 1969. First published in 1948. Covers in delightful detail mechanical inventions for daily living, particularly for the period from 1850 on. Among other things, the book manages to make the mechanization of the bathroom fascinating.

Goldstine, Herman H. *The Computer from Pascal to von Neumann*. Princeton: Princeton University Press, 1972. A richly detailed account of the development of computers, particularly in the crucial 1940s.

Grant, Michael. *The Twelve Caesars*. London: Michael Grant Publications, 1975.

Greenberger, Martin, ed. *Computers and the World of the Future*. Cambridge, Mass.: MIT Press, 1962.

Hadas, Moses. *Ancilla to Classical Reading*. New York: Columbia University Press, 1954. For background on ancient techniques of writing and oral reading.

Hamming, R. W. *Introduction to Applied Numerical Analysis*. New York: McGraw-Hill, 1971. A textbook by one of the leading computer mathematicians.

Herder, Johann Gottfried von. *J. G. Herder on Social and Political Culture*. Edited by F. M. Barnard. Cambridge: Cambridge University Press, 1969. Contains Herder's "Essay on the Origin of Language," which joins in the eighteenth-century debate about "artificial" language.

Hofstadter, Douglas R. *Gödel, Escher, Bach: An Eternal Golden Braid*. New York: Basic Books, 1979. A book that expresses perfectly the spirit of Turing's man. Hofstadter may well be defining many of the "philosophical" problems that will occupy our culture for years to come: the paradox of meaning within meaningless formal systems, self-reference, finite and infinite loops, artificial intelligence, the mind as a semantic network, and so on. The author's attitude toward art—as a play of formal elements without any true historical dimension—also reveals how remote Turing's man is from his predecessor.

Hunt, Earl. "What Kind of a Computer is Man?" *Cognitive Psychology* 2 (1971): 57–98. Another explicit acceptance of the metaphor of electronic man.

Innis, Harold A. *Empire and Communications*. Toronto: University of Toronto Press, 1972. Originally published in 1950. Drops any number of fascinating hints about time, space, language, and culture, but in a disconcerting, telegraphic style.

Jammer, Max. *Concepts of Space*. Cambridge, Mass.: Harvard University Press, 1969. A general treatment from Aristotle to Einstein.

Contains a canny suggestion that Einstein's "spherical space" is in some sense a return to Aristotle's finite universe (p. 22).

Kemeny, John G. *Man and the Computer*. New York: Charles Scribner's Sons, 1972. Lectures delivered by a computer specialist and educator, the inventor of BASIC.

Kidder, Tracy. *The Soul of a New Machine*. Boston: Little, Brown and Company, 1981. Examines the psychology of a design team working on a new minicomputer; a valuable look at how such designers think and work.

Kirk, G. S., and Raven, J. E. *The Pre-Socratic Philosophers*. Cambridge: Cambridge University Press, 1957. Fragments of the philosophers with translation and analysis. I have used in particular the discussion of the Pythagorean attitude toward number, which was so influential in the ancient world.

Knuth, Donald E. *Mathematics and Computer Science: Coping with Finiteness* (Stan-CS-76–541). Stanford: Stanford University, Computer Science Department, February 1976. A lecture, in the author's usual playful style, on the important idea that the computer can only deal with finite numbers.

Koyré, Alexandre. *From the Closed World to the Infinite Universe*. Baltimore: Johns Hopkins University Press, 1974. Chronicles this great change in Western thinking, principally through the sixteenth, seventeenth, and eighteenth centuries.

Kuhn, Thomas S. *The Structure of Scientific Revolutions*. Chicago: University of Chicago Press, 1962. Probably belongs in any contemporary bibliography on a scientific subject.

Leibniz, Gottfried Wilhelm von. *The Philosophical Writings of Leibniz*. Selected and translated by Mary Morris. London: Everyman's Library, 1934.

Levy, David. *1975 US Computer Chess Championship*. Woodland Hills, Calif.: Computer Science Press, 1976.

Lindsay, Peter H., and Norman, D. A. *Human Information Processing*. New York: Academic Press, 1972. Another example of the borrowing of computer concepts by psychologists.

Lloyd, G. E. R. *Aristotle: The Growth and Structure of His Thought*. London: Cambridge University Press, 1968. A good introduction.

McCorduck, Pamela. *Machines Who Think*. San Francisco: W. H. Freeman, 1979. The authorized court history of artificial intelligence and, as such, a fascinating look into the motivations and aspirations of these computer specialists.

Macey, Samuel L. *Clocks and the Cosmos: Time in Western Life and Thought*. Hamden, Conn.: Archon Books, 1980. Numerous examples of the clock metaphor from philosophy, literature, and art. Focuses on the seventeenth and eighteenth centuries.

McLuhan, Marshall. *The Gutenberg Galaxy: The Making of Ty-

pographic Man. Toronto: University of Toronto Press, 1972. The printing press and cultural change. Whatever one thinks of McLuhan's public persona, one must admit that he was in possession of a very important idea.

Mead, Carver, and Conway, Lynn. *Introduction to VLSI Systems.* Reading, Mass.: Addison-Wesley, 1980. An excellent textbook explaining the design and fabrication of microcomputer systems.

Miller, G. A. "The Magical Number Seven, Plus or Minus Two: Some Limits on Our Capacity for Processing Information." *Psychological Review* 63 (1956): 81–97. An early and influential piece on man as an information processor; human memory as storage device.

Minsky, Marvin. *Computation: Finite and Infinite Machines.* Englewood Cliffs, N.J.: Prentice-Hall, 1967.

——————. "Steps Toward Artificial Intelligence." *Proceedings of the IRE* 49 (1961): 8–30.

——————, ed. *Semantic Information Processing.* Cambridge, Mass.: MIT Press, 1968. A collection by artificial intelligence advocates.

Morrison, Philip, and Morrison, Emily, eds. *Charles Babbage and His Calculating Engines.* New York: Dover Publications, 1961. A fascinating collection of excerpts and papers by Babbage himself and his followers.

Mumford, Lewis. *The City in History.* New York: Harcourt, Brace & World, 1961.

——————. *Myth of the Machine.* Vol. 1, *Technics and Human Development.* New York: Harcourt, Brace & World, 1967. Vol. 2, *The Pentagon of Power.* New York: Harcourt Brace Jovanovich, 1970. This late work takes a negative view of our technological future.

——————. *Technics and Civilization.* New York: Harcourt Brace, 1934. A brilliant work showing the intimate connection between technology and culture from the Middle Ages to the twentieth century.

Nelson, Theodor H. *Computer Lib.* South Bend, Ind.: Theodor H. Nelson, 1974. A book that is determined to "popularize" the computer at any cost; still, the author has recognized some of the key qualities of the machine.

Newton, Isaac. *Sir Isaac Newton's Mathematical Principles of Natural Philosophy and His System of the World.* 2 vols. Translated by Andrew Motte in 1729. Revised and edited by Florian Cajori. Berkeley: University of California Press, 1966.

Ong, Walter J. *Ramus, Method, and the Decay of Dialogue.* Cambridge Mass.: Harvard University Press, 1958.

Oresme, Nicole. *Le Livre du ciel et du monde.* Translated by A. D. Menut. Edited by A. D. Menut and A. J. Denomy. Madison: University of Wisconsin Press, 1968.

Papert, Seymour. *Mindstorms: Children, Computers, and Powerful Ideas*. New York: Basic Books, 1980. An intriguing combination of educational theory and computer programming. Papert has created a programming language through which children can learn the procedural thinking of the machine.

Pennington, Ralph H. *Introductory Computer Methods and Numerical Analysis*. New York: Macmillan Co., 1965.

Pizer, Stephen M. *Numerical Computing and Mathematical Analysis*. Chicago: Scientific Research Associates, 1975.

Price, Derek J. de Solla. "An Ancient Greek Computer." *Scientific American* (June 1959): 60–67. Describes the paradoxical Antikythera device.

Ralston, Anthony, and Meek, C. L. *Encyclopedia of Computer Science*. New York: Petrocelli/Charter, 1976. A good reference work for the nonspecialist.

Raphael, Bertram. *The Thinking Computer: Mind Inside Matter*. San Francisco: W. H. Freeman, 1976. By a proponent of artificial intelligence.

Robins, R. H. *A Short History of Linguistics*. Bloomington: Indiana University Press, 1967. A well-balanced presentation, from the Greeks to the 1960s.

Russell, Bertrand. *Introduction to Mathematical Philosophy*. London: George Allen and Unwin, 1967.

Sagan, Carl. *The Dragons of Eden: Speculations on the Evolution of Human Intelligence*. New York: Ballantine Books, 1977. Agrees with many in seeing computers as a further evolution of the human brain.

Simon, Herbert. *The Sciences of the Artificial*. Cambridge, Mass.: MIT Press, 1969. Perhaps the most famous proponent of artificial intelligence, Simon wholeheartedly endorses the notion of man as information processor.

Singer, Charles J., et al. *A History of Technology*. 7 vols. Oxford: Clarendon Press, 1954–78. An encyclopedic collection of rather long chapters.

Solmsen, Frederick. *Aristotle's System of the Physical World*. Ithaca: Cornell University Press, 1960. A masterful explanation of the development of Aristotle's thinking out of and away from Plato's.

Spengler, Oswald. *The Decline of the West*. 2 vols. Translated by C. F. Atkinson. New York: Alfred A. Knopf, 1945. Much maligned but brilliant in its way. Among other accomplishments, Spengler understood the meaning of limit for Greek culture and infinity for the West. If I am right, a returning appreciation of the finite will be a defining quality of the computer age.

Taube, Mortimer. *Computers and Common Sense: The Myth of Thinking Machines*. New York: Columbia University Press, 1961. Against artificial intelligence.

Timmerman, Peter. *Vulnerability, Resilience, and Society*. Toronto: University of Toronto, Institute for Environmental Studies, 1981.

Turbayne, Colin M. *The Myth of Metaphor*. New Haven: Yale University Press, 1962. A philosophical essay on the place of metaphor in science.

Turing, A. M. "On Computable Numbers, with an Application to the Entscheidungsproblem." In *Proceedings of the London Mathematics Society*, 2d ser., 42:230–65. London: C. F. Hodgson, 1936. The paper in which Turing defined his logical machine.

Vartanian, Aram. *La Mettrie's L'Homme-Machine: A Study in the Origins of an Idea*. Princeton: Princeton University Press, 1960. An essay tracing the idea from Descartes's bête-machine to La Mettrie, together with a full translation of La Mettrie's text.

von Neumann, John. *Collected Works*. Edited by A. H. Taube. 6 vols. Oxford: Pergamon Press, 1961–63. Contains von Neumann's well-known papers on computers and the theory of automata.

——————. *The Computer and the Brain*. New Haven: Yale University Press, 1958. Analyzes the brain in the operational terms supplied by the digital computer.

Weizenbaum, Joseph. *Computer Power and Human Reason*. San Francisco: W. H. Freeman, 1976. One of the few attempts to deal with the cultural impact of the computer. Weizenbaum is an opponent of artificial intelligence.

White, Lynn T. *Medieval Technology and Social Change*. Oxford: Clarendon Press, 1962. Exciting chapters on cavalry, agriculture, and mechanical-dynamic technology.

Whitehead, A. N. *Science and the Modern World*. New York: Macmillan Co., 1967.

Whorf, Benjamin L. *Language, Thought, and Reality*. Edited by J. B. Carroll. Cambridge, Mass.: MIT Press, 1956. Often cited as the source of the idea that our native language colors our whole world view.

Wickelgren, Wayne A. *Learning and Memory*. Englewood Cliffs, N.J.: Prentice-Hall, 1977. I cite this book as an example of the extent to which computer jargon has been taken over into cognitive psychology.

Wiener, Norbert. *Cybernetics*. New York: John Wiley and Sons, 1948. An elegant account of the identification of man and machine by the famous mathematician and student of technology. Wiener is in many ways a more civilized and sympathetic proponent of the making of electronic man than current writers.

Wilkins, John. *An Essay Towards a Real Character and a Philosophical Language*. London, 1668. Reprint, edited by R. C. Alston. English Linguistics 1500–1800, no 119. Menston, Eng.: Scolar Press, 1968. A beautiful book that presents calligraphic symbols to represent basic ideas. It is representative of the general movement for a logical language of thought, as Leibniz was suggesting.

Winograd, Terry. *Understanding Natural Language*. New York: Academic Press, 1972. A description of SHRDLU, the simulated robot. Very influential for the artificial intelligence movement.

Yates, Frances. *The Art of Memory*. Chicago: University of Chicago Press, 1966. Wonderfully detailed account of ancient, medieval, and Renaissance mnemonic systems.

Index

Address, 83–85, 152, 155, 198, 241; converted by assembler, 128. *See also* Random access; Linear access

Address space. *See* Logical space

Algorithm, 52, 59, 64, 127, 130, 168, 176, 244; as analogy for human thought, 13, 151, 194, 200, 221, 233, 241

Analytical Engine. *See* Babbage, Charles

Antikythera device, 15

Arabs: and mechanical clocks and devices, 16, 24; and algebra, 60

Architecture, Greek: use of space in, 80, 93, 96

Architecture, Roman, 63; aqueducts, 18, 24

Aristotelians, 94

Aristotle, 74, 80, 92, 93, 124, 136, 141, 181, 216; and form and matter, 17, 23, 182; dislike of infinity, 63; and logic, 69, 73, 136, 137; *Nicomachean Ethics*, 73; world as a finite plenum, 91, 95, 97; heavenly spheres, 116, 182

Arithmetic. *See* Computer mathematics; Mathematics

Artificial intelligence, 12, 13, 52, 143, 152, 189–213, 219, 220, 221, 233–35, 238, 241

Artificial language, 142, 146–148, 172, 173, 244; compared to natural language, 124–127; hierarchical character of, 127–132. *See also* Assembly language and assemblers; Machine language; Programming language

Assembly language and assemblers, 128, 129, 130, 197, 236, 241, 242, 243. *See also* Machine language; Programming language

Automaton: automata as imitations of life, 41, 203–206, 208, 212–13, 232; Turing machine, 114, 186. *See also* Hero of Alexandria; Jacquet-Droz; SHRDLU; Turing machine; Vaucanson, Jacques de

Automobile, 5, 49, 90, 110, 120, 121, 139, 200

Babbage, Charles, 66; his Difference Engine, 32; his Analytical Engine, 32, 33, 53, 66, 69, 161, 177; on infinity, 177

Bacon, Francis, 119, 139

Bruno, Giordano, 204; and the art of memory, 160, 164

Byte, 82, 83, 84, 85, 111, 152, 153, 154, 175, 179, 242, 243, 245, 246

Calculus, 58, 60, 61, 64, 142; logical calculus, 49, 69, 73, 121, 142–44, 199, 241. *See also* Computer mathematics; Mathematics; Numerical analysis

Camillo, Giulio: his memory theater, 159, 160, 164

Capitalism, 123

Central processing unit: characteristics and operation of, 34, 35 (figure 2-3), 38, 42, 48–49, 50–51 (figure 3-2), 52, 67, 68, 71, 81, 82, 83, 127–28, 129, 131, 146, 151, 152, 153, 154, 157, 165, 167, 168, 169, 178, 179, 200, 227, 241, 242, 243, 244, 245, 246; and computer mathematics, 54, 64; and qualities of logical thought, 75–79; as electronic clock, 101–105, 109, 110, 111, 112, 115. *See also* Von Neumann computer

Charlemagne: received water clock, 24

Checkers: played by computer, 199